T0073047

Understanding and Changing the World

Joseph Sifakis

Understanding and Changing the World

From Information to Knowledge and Intelligence

Joseph Sifakis
Verimag Laboratory
Université Grenoble Alpes
St Martin d'Hères, France

ISBN 978-981-19-1998-5 ISBN 978-981-19-1932-9 (eBook)
https://doi.org/10.1007/978-981-19-1932-9

This Springer imprint is published by the registered company Springer Nature Singapore Pte Ltd.
The registered company address is: 152 Beach Road, #21-01/04 Gateway East, Singapore 189721, Singapore

Πάντες ἄνθρωποι τοῦ εἰδέναι ὀρέγονται φύσει—All men naturally desire knowledge.
Aristotle, Metaphysics

Preface

I would like to briefly discuss the motives that have prompted me to write a book with such a broad and ambitious objective that goes beyond the scope of my specific scientific and technical knowledge.

In the early 1980s, I came to realize that my research into applied logic and mathematical language theories was directly related to philosophical problems, particularly the problem of consciousness and language.

I thus started reading and became interested in philosophy. I must say that, despite my best efforts, I could not find a path through the bewildering maze of philosophical theories. It bothered me that many philosophers are not strict enough in their thinking. Their discourse is nearer to the literary. This may well be due to certain individuals who happen to be fine wordsmiths, who know how to write elegant and exciting text. However, most philosophers did not convince me, as I am possessed of the strict, pedestrian mind of an engineer. I find that many play around with terminology, creating new terms without actually caring about how they relate to existing ones. What has particularly struck me is the existence of closed communities around a philosophical or even scientific trend. Organized by a clergy of "authorities," they use hermetic terminologies that render their theories inaccessible to the "uninitiated" and lends them legitimacy in the eyes of laymen. Thus, there are existentialists, structuralists, empiricists, Marxists, Hegelians, etc., each philosophizing from their own context and pulpit, clinging to their "truths" and disregarding everyone else, without a thought to how their knowledge helps people live a life worth living. This may explain why the present day is characterized by a depreciation of philosophy and the humanities, by a dearth of philosophical education.

I believe that any attempt to formulate our knowledge must respect the laws of an economy of thought, must be based on a set of well-founded concepts that enjoy clearly defined semantic relationships with each other.

After much philosophical perambulation and searching, I decided in the late 1990s to strictly formulate my views on the world starting with what I am sure and aware of. The truth, as it is created and processed by the human mind, cannot by its very nature be complex. All theories and knowledge of any utility are shockingly simple in their principles—regardless of whether we use specialized and sophisticated techniques to study them. I believe that it is dangerous for our societies to become alienated from scientific and technical knowledge. We must give them the tools to judge, at the very least, its importance and the impact of its application.

I wanted to gradually build an outlook of the world where I always distinguish what I can talk about with some degree of certainty from what I do not know. Our aim must be to push the boundaries of knowledge as far as possible, while remaining aware of our limitations.

In my searching, I quickly identified the issues of knowledge and, therefore, consciousness and language as the key themes.

For decades, I have been grappling with human intelligence, whether it is possible for a computer to approximate it. Can one study human behavior—while avoiding the traps of simplification—and draw certain useful, practical conclusions, for example, on what freedom of will is and what its management means, what a deadlock and what its solution consists of?

I reached the conclusion that the world is the combination of two distinct, yet complementary entities: matter/energy on the one hand and information on the other—what can be seen and what can be thought.

I despise moralism and vagueness. I believe that even ethical problems have a technical dimension that we must understand in order to live better lives having, naturally, first identified the limits of such an approach.

What has prompted my investigations and inquiries is the realization that we are living in a time of absolute relativism and confusion. The crisis we have been living through for decades has gravely necessitated a philosophy of the individual. One that explains in a simple and comprehensible manner, free of dogmatism, the practical necessity of values and ethics for individual happiness and social peace and progress. One that helps us understand the importance and limits of knowledge, deepen our self-knowledge, and delve into the dual-nature, double-faceted game of comprehension: conscious and subconscious, logical and intuitive, empirical and transcendental.

It is shocking how real problems are being thoroughly hidden from the public and how the conditions for concealing them are artificially created. At the same time, these real problems are replaced by other fictitious ones, which disorient people and divert them from the conscious search for solutions. Scientific and technical progress is useless, if not dangerous, for societies unworthy of managing and directing it. Even the most perfect legal system is ineffectual without free compliance with ethical rules. Lack of intellectual values cannot be offset by tangible goods.

Over the course of my research and academic career spanning 50 years, my work has focused on building cost-effective, trustworthy computerized systems. My early work on systems verification met with broad application in systems engineering. I later studied system design as a process that leads from technical requirements to artifacts satisfying them. This led me to investigate the role of knowledge in problem-solving and its inherent complexity limitations. Over the past 10 years, I have worked on the design of autonomous systems and of self-driving cars in particular. My research on autonomous systems design has deeply informed and inspired the ideas about intelligence and consciousness developed in the book.

I believe that the criterion for intelligence is not just a Q&A game. It is about building systems that can replace human agents performing tasks in complex organizations. Therefore, apart from solving hard computational problems, it also involves challenging systems engineering issues that have been often underestimated.

It was my wish and responsibility to write about the importance of knowledge as an intangible entity, separate from physical entities, and the need to understand its development and implementation processes; to show how new concepts and models stemming from informatics can help us deepen the functioning of individual and collective consciousness; and to demonstrate how all the structures that make up our societies implement above all else information systems whose proper functioning depends on the seamless and timely exchange of reliable knowledge.

It is my hope that this endeavor marks the first step in the reader's introduction to the concept of knowledge, its multiple forms and uses, that it enables readers to see the world through a gnoseological lens and empowers them to act properly, both as individuals and as citizens.

My ideas have been influenced by numerous texts as well as exchanges of views with colleagues and good friends. I am grateful to Cristian Calude for reading the manuscript and for his insightful and constructive remarks.

I would like to acknowledge the role of Giorgos Hatziiakovou, Director of Armos Publications, who believed in the book from the very outset and published its first version in Greek. I would also like to heartily thank Daniel Webber who translated a revised version of the Greek text into English with a lot of competence, diligence, and talent.

Finally, I could not have written this book without the constant, undivided support of my wife, Olga. Her love of all that is elevated and wonderful, spiritual communion and shared experiences were catalytic in inspiring me to clarify and formulate core ideas in the book.

Meylan, France Joseph Sifakis

Contents

Chapter 1
Introduction

The book focuses on knowledge as useful and valid information allowing us either to understand the world or to change the world in order to satisfy material and spiritual needs. It provides a thorough treatment of knowledge in all its aspects and applications by machines and humans, free of ideological and philosophical bias and relying exclusively on concepts and principles from the theory of computing and logic. As such, it strikes a balance between popular books skirting fundamental issues and often focusing on the sensational, and scientific/technical or philosophical books that are not accessible to laymen.

The book progressively introduces the concept of information as an intangible entity developed and used by both computers and living organisms. It takes a unified approach considering the many facets of knowledge depending on their degree of validity and truthfulness; these include implicit empirical, scientific and technical, mathematical knowledge as well as meta-knowledge used to combine the different types of knowledge in order to solve problems. The book also considers the different mechanisms and processes for developing knowledge and their interrelationships by examining their main characteristics and limitations. This provides a basis for a discussion about the concept of intelligence and a comparison between human and artificial intelligence (AI).

The book attempts an analysis of technical difficulties for reaching human-level intelligence from weak AI that currently provides only intelligent systems able to solve specific problems. It argues that the strong AI vision is not reachable by incremental improvement of machine learning technology. The development of trustworthy autonomous systems, as anticipated by the Industrial Internet of Things, is considered a bold step toward closing the gap between human and machine intelligence.

The book discusses a characterization of autonomous behavior in terms of five basic functions and shows how this model can be adequately extended to capture key features of human consciousness. This leads to a computational scheme that could explain conscious behavior as a game where decision-making seeks tradeoffs

J. Sifakis, *Understanding and Changing the World*,
https://doi.org/10.1007/978-981-19-1932-9_1

between satisfaction of needs and the corresponding costs estimated using an individual value system.

Finally, the book extends the model of individual consciousness into a model of collective consciousness shaped by the action of institutions fostering a common value system. It corroborates the thesis that human societies are information systems, illustrated by the analysis of the underlying organizational principles as well as of crises as manifestations of failure in their application.

The book differs from others on the philosophy of science that focus on epistemology (from the Greek "episteme"—science) examining the human-driven process for acquisition of knowledge about the world. It takes a broader perspective considering all kinds of knowledge. This is why, instead, I use the term gnoseology (from the Greek "gnosis"—knowledge), which is more appropriate, since it suggests that knowledge is not limited to the understanding of phenomena by humans, as discussed in the book.

The approach taken clearly separates physical reality from its models. It follows the dualist tradition of logicians emphasizing the importance of logic and language. In particular, the concept of information being considered is purely logical and different from "information" in quantum physics or in the theory of communication. From this perspective, the book mounts some hopefully well-founded criticism toward approaches considering that information and knowledge are just emergent properties of physical phenomena.

I have tried to explain my ideas as simply and clearly as possible while avoiding dangerous oversimplifications. Where necessary, the text veers toward somewhat technical, but without the knowledge required to exceed the 12th-grade level. These small technical asides can be skipped during a first reading.

Sometimes, my love for ancient Greek literature comes through the exposition, particularly with references and citations. I hope that this would be of interest to international readers and would allow them insight into the sinuous course of the human spirit in the quest for knowledge.

The scope of the issues I am discussing is such that I do not think it expedient to provide bibliographic sources. I make certain references to websites, in particular Wikipedia, which provide additional information and, above all, relevant bibliographic sources.

The book is composed of three parts.

The first part, consisting of three chapters, presents knowledge as truthful and useful information, as well as the processes of its development and management by computers and humans.

The second chapter identifies three fundamental questions about knowledge: teleological, ontological, and gnoseological. We explain that only gnoseological questions can be approached strictly from a methodological and logical point of view.

In the third chapter, we outline the basic concepts of informatics. We explain, in the simplest possible terms, how knowledge is defined as information, and we discuss key problems posed by its development and application. In particular, we endeavor to analyze the links between scientific and technical knowledge.

In the fourth chapter, we discuss principles of knowledge development that are based on simplifications of reality, as well as restrictions arising from these principles.

The second part, consisting of two chapters, analyzes the relationship between computational processes and physical phenomena, as well as the production and application of knowledge by machines and humans.

The fifth chapter investigates relationships between informatics and physics and, in particular, discusses how each of these fields of knowledge can enrich the other, comparing basic concepts and models.

The sixth chapter attempts to compare artificial and natural intelligence, trying to answer the question of whether and to what extent computers, as they stand today, can approach human mental functions. The chapter presents autonomous systems that are called upon to replace humans in complex operations by realizing the vision of strong AI. Finally, it discusses the risks—real or hypothetical—of the reckless use of computers and artificial intelligence.

The third part, consisting of two chapters, seeks to analyze the functioning of individual consciousness and to discuss its impact on societal organization, in the light of the previous gnoseological perspective.

The seventh chapter explains the functioning of consciousness as an autonomous system that manages short- and long-term goals based on value criteria and accumulated knowledge.

The eighth chapter discusses how values are shaped in society and how the composition of each individual's subjective experience takes on an objective dimension, which can be studied as a social phenomenon. It analyzes the role of institutions in shaping and maintaining a common scale of values for strengthening social cohesion, as well as a discussion on the principles of democracy.

The book concludes with an epilogue summarizing key conclusions and questions on what the future holds for us.

Part I
For a Gnoseological View of the World

We present knowledge as truthful and useful information, as well as the processes of its development and management by computers and humans.

- We identify three fundamental questions about knowledge: teleological, ontological and gnoseological. In it, we explain that only gnoseological questions can be approached strictly from a methodological and logical point of view.
- We outline the basic concepts of informatics and explain how knowledge is defined as information.
- We discuss principles of knowledge development that are based on simplifications of reality, as well as restrictions arising from these principles.

Chapter 2
Fundamental Questions About Knowledge

2.1 Three Key Questions—Why? What? How?

Our efforts to understand the world raise three kinds of fundamental questions about knowledge.

First, *teleological* questions, seeking a purpose in phenomena such as "why did the world form?" and "why do we exist?". The use of adverb "why" implies will and intent, which are traits of human consciousness. Therefore, the answers to such questions cannot be controlled through reason and are beyond any experimental validation and analysis.

Second, *ontological* questions, which concern the nature of "being." For example, "does the world exist outside our senses and our minds?" or "what precisely is the world as such and as a synthesis of its elements?".

These are questions about *what* we can know. We ask such questions because we do not distinguish between reality and its models. These are conceptual abstractions our minds use to understand reality. Nowadays, with quantum mechanics and computers, it is commonly accepted that the world and its models are two separate things. I will explain below why ontological questions are not amenable to rational control either.

Third, *gnoseological* questions about *how* we can know, for example, how the world is changing, how we think, how we build a building, and how birds fly. Answers to these questions allow us to understand or to change the world. Knowledge can be scientific, technical, mathematical, or purely empirical, as I will explain in what follows. It consists of relationships that we can verify logically or validate empirically using the appropriate methodology.

This simple classification clearly and strictly establishes a typology of fundamental questions about knowledge. As regards teleological and ontological questions, we do not have any logical or empirical criteria to decide whether or not an answer is sound. It is only worth attempting to answer gnoseological questions, insofar as they are formulated as relationships between well-defined concepts.

J. Sifakis, *Understanding and Changing the World*,
https://doi.org/10.1007/978-981-19-1932-9_2

It should be noted that these three types of questions are independent from each other. Whatever your convictions on "why?" and "what?", they do not logically affect your knowledge about the world. Knowledge does not depend on whether you are a Buddhist or an atheist, whether or not you believe numbers exist regardless of the human mind.

However, to avoid misunderstanding, I must stress that I do not disregard the importance of teleological and ontological problems in the philosophy of the individual. It is simply that the answers each person adopts are derived from their convictions and cannot be judged according to objective criteria.

Disputes between religion and science arise from time to time when limits are violated, i.e., when religion asserts opinions on gnoseological issues, for example, disputing heliocentrism or when science encroaches into metaphysical territory without taking the necessary precautions. Certain scientists overlook inherent limitations of scientific theories and the relativity of scientific knowledge or use scientific methodology in an abusive and inappropriate manner, as I will discuss in what follows.

2.2 The Teleological "Why?"

The most difficult questions are the "why?"s that torment us from a very young age. "Why does the world exist?", "why are we here?", "why do we come and go?", "why are we conscious?", and a host of others you can imagine. All of these questions are linked to seeking purposes. We are searching for interrelationships in everything we see and analyze. And even if we do answer a "why?", for example, "why do we die?", other, deeper "why?"s emerge. This creates chains, which result in teleological questions that cannot be answered rationally.

Plato accepts that nature was designed by an intelligent creator according to pre-existing immutable models. The idea of teleology starts with Aristotle, on whom Christian theologians/philosophers later relied. Aristotle criticizes Democritus for his overemphasizing the necessity of physical laws but without identifying any "final purpose"—"it is absurd to say that there is no purpose in nature simply because we do not see it."

It is interesting that, through historical materialism, Marx and Engels adopt a sort of "teleological" outlook on sociohistorical processes, the difference being that they replace the mind and consciousness with economic determinism.

The ancient Greeks took the world as it is and their questions were more of "what?" and "how" varieties. Subsequent teleological philosophical ideas were influenced by Judaism and Christianity, dominated by the idea that God has a plan for humanity. Perhaps it is no accident that the term "project" does not have a precise equivalent in the Greek language, one so rich in concepts.

In a strictly logical approach, teleological ideas boil down to rhetoric, to a mythology that does not affect the way knowledge is sought. Of course, nothing prevents us from teleologically interpreting the laws governing phenomena, but this is logically superfluous. Thus, some people may ascribe a higher purpose to the law of conservation of mass/energy, but this cannot have any rational basis.

2.3 The Ontological "What?"

Parmenides may have been the first thinker to raise the question regarding "being" and its importance. The Parthenon is a "being," it exists. However, I perceive it in one way, a foreign visitor in another, an architect in another, and an archeologist in yet another. Yet, where does its "being" lie? The fact that it is made of marble, in its overall form and architecture, in what it depicts?

Parmenides would reject the sensible world: "he made reason the standard, and pronounced sensations to be inexact." He would accept as what "exists" only what is recognized and determined by "reason" and identified with it.

To put it another way: understanding reality is reduced to a process of interpretation carried out by the mind. This point of view relativizes knowledge and sets the "substance" of reality beyond its scope. At the same time, though, it resolves certain false problems encountered by those who have attempted such an endeavor.

The fact that each mind can see reality differently does not negate the existence of commonly accepted views on reality. This is achieved through communication, the formation of shared knowledge through social interaction.

The quest for an absolute substance of things and pure categories of "being" has troubled and continues to trouble scientific and philosophical thinking. Aristotle accused Heraclitus of being irrational for blatantly ignoring the basic logical principle of "contradiction" and the "excluded middle."

To take an example from the more recent past, let us examine the dispute around the particle-or-wave nature of the electron that shook up twentieth-century physics. Many major thinkers saw a clear logical contradiction in the fact that the electron "is simultaneously" both a particle (discrete) and a wave (continuous). Heisenberg, like many others, would reach the hasty conclusion that "the mathematical scheme of quantum theory can be interpreted as an extension or modification of classical logic," meaning a logic where the principle of contradiction does not hold true.

However, a Parmenidean view of reality rids us of these impasses and paradoxes—as long as we give up any ambition of understanding what an electron "truly is," of course. Contradictions are resolved if we understand that, depending on how we observe them, phenomena can be consistent with models that have different characteristics, without this necessarily implying a logical contradiction. Reality is "what it is" and we can only mentally understand reality through theories that abstract it. The multiplicity of theories does not necessarily entail logical contradictions, provided each theory takes different aspects of reality into account. Of course, when two or more theories are applicable, an interesting problem that arises concerns the study of the relationships between them—which was successfully done by quantum physics.

I believe that, in this way, one can understand Heraclitus's thinking, which proposes a dialectical perception of reality. The harmonious synthesis of opposites does not lead to a logical contradiction. It is a paradigm for understanding reality not as something static, but as an "evolving process." Evolution, motion, and everything that changes is the result of the synthesis of counteracting forces that shift their point of equilibrium according to their balance.

2.4 The Gnoseological "How?"

The type of questions open to rational investigation is those asking "how?". How does a clam reproduce, how is force related to acceleration, how do you build a house, how do you solve a mathematical problem, how do we avoid inflation? The answers to these questions are relationships between observations involving concepts and objects. They comprise what I call *knowledge*, which I will define more precisely in what follows.

At this point, I should stress that these questions concern both understanding situations and solving problems. Moreover, they concern not just the physical world, but also abstract mental processes.

Knowledge is about questions open to answers that are general and independent of each individual's personal experience. It must be possible to affirm their truth in an empirical or logical way. That is why some people believe that one can only know the physical world and what they call "objective reality." Such a view implies that mental phenomena and consciousness can only be understood to the extent that they can be studied based on the laws of physics. This often leads to the position that the world is divided, on the one hand, into physical phenomena that are a "game" of particles in fields of forces and, on the other hand, into the world of perceptions, of consciousness, which is the subjective experience of each individual.

Here I agree with the philosopher John Rogers Searle, who points out that such a distinction leads to a false dichotomy. Ontologically speaking, consciousness is certainly a subjective experience. Death or pain are ontologically subjective experiences, but gnoseologically objective, insofar as they can be ascertained in an unequivocal way. Thus, it makes sense to study the causes of a person's death or the causes and treatments of pain.

A key tenet of this book is that knowledge is not just about physical reality but also about all mental and social phenomena. As I will explain in what follows, this implies the dual nature of reality, which consists of physical phenomena and informational phenomena.

Chapter 3
Information and Knowledge

3.1 The Birth of Informatics

Informatics is a new field of knowledge. It was established in 1936 by British mathematician Alan Turing, who defined a mathematical model of computation—the machine bearing his name.

The rest is, more or less, history. Thanks to rapid advances in electronics and materials technology, we developed increasingly powerful computers. The first computer was built for military applications. This was followed by the use of computers for commercial applications around the 1960s. We then connected computers to telecommunications networks, and this paved the way for the Internet and the Web—the information society. Another important step was the use of computers in embedded systems to control production processes and develop services. Today, over 95% of integrated circuits are embedded. Concealed in devices of all kinds, they provide automated services.

At first, informatics was considered to be a technology lying somewhere between mathematics and electronics. Over the years, it developed as an independent field of knowledge and resisted absorption by other fields. Today, everyone agrees that it is an important field of knowledge that extends far beyond "computer science." "Computer Science is no more about computers than astronomy is about telescopes," as Edsger W. Dijkstra used to say [1].

What we have been living through the last 60 years, the information revolution and, by extension, the knowledge revolution, is one of the most important milestones in the history of humanity.

Since antiquity, people have been building machines to multiply their muscular power, to set up the technical civilization that has so radically changed the world. Thanks to computers, people are able to staggeringly multiply their mental abilities. This is because computers far surpass humans in terms of speed and accuracy of computation, while creativity and the ability to understand the world remain the privilege of humans. There is, thus, considerable complementarity.

J. Sifakis, *Understanding and Changing the World*,
https://doi.org/10.1007/978-981-19-1932-9_3

We are at a great landmark in humanity's path toward conquering knowledge, which began with the appearance of language, the discovery of writing, and the invention of printing.

Despite its importance, informatics has not been accordingly acknowledged in primary and secondary education. It does not receive the same emphasis as physics or biology. It is taught more as a form of technology, without students delving into theories of computation. Public opinion and decision-makers do not even grasp its importance as a field of knowledge that can fundamentally change not only our perception of the world but also the very existence of human beings.

The truth is that the so-called exact sciences dominated scientific thinking until the end of the twentieth century. As an academic, I had the opportunity to converse with famous scientists and was often surprised to realize a lack of understanding of the importance of informatics. It is true that scientists try to explain the world and everything complex as a combinatorial game of particles, the whole as a synthesis of its parts. This reductionist approach wants complex phenomena to "emerge" from the synthesis of simpler ones.

As Richard Dawkins characteristically notes [2]: "My task is to explain elephants, and the world of complex things, in terms of the simple things that physicists either understand, or are working on." However, this approach fails to explain what information is and what computation is. At times, it leads to ridiculous assertions, such as Stephen Hawking's claim that "the brain could exist outside the body" [3]. In other words, he believes that he could go on living after death by placing his brain's information in a memory stick.

Such statements suggest a deep ignorance of what information is, of how cognition and consciousness work. It is a pity that they strike a significant chord with the media.

I would like to explain as simply as possible what the nature of information and computation is. Needless to say that at present, there is still no full agreement on these concepts, even among computer scientists.

The debate on the relationship between informatics and other fields of knowledge raises deep and delicate epistemological problems. My aim is not to provide answers to these problems, which will remain unresolved for a long time. I simply propose a methodological framework for comparing informatics to other fields of knowledge. I will also explain what can be achieved with computers today, what their inherent limitations are, and how much they can contribute toward developing fruitful interdisciplinary cooperation.

3.2 What Is Information?

Informatics involves the study of computation as a process for transforming information—what we can compute and how to go about computing it.

Yet, what is information?

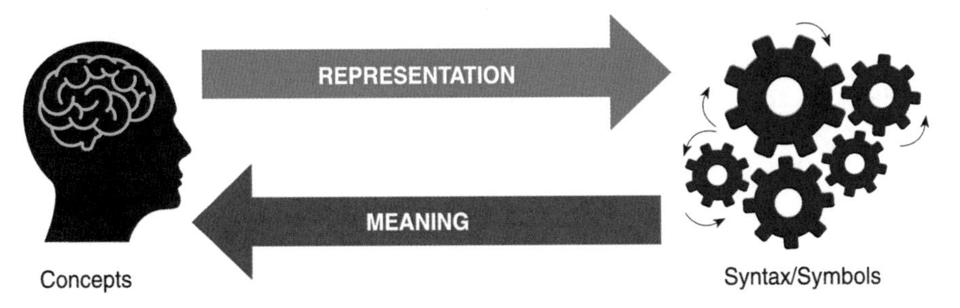

Fig. 3.1 Information as a relationship between symbols and concepts

Let me first give an example. When I see the symbols 4, 100, δ, IV, I can interpret them as the concept of "four" in the decimal system, in the binary system, or the ancient Greek and Roman numeral system, respectively. What the symbols represent is a matter of convention. The interpretation of the information codified in these symbols depends on whether the mind of the person seeing the symbols is familiar with this convention. Familiarization can take place either through empirical learning or through education. For all non-color-blind people, the color red corresponds to the concept "red," and the photograph of an apple recalls the concept of an "apple." Conversely, Maxwell's equations constitute information only for those familiar with electromagnetism. The Linear A script, insofar as it has not been deciphered, does not convey information.

Information can be defined as a semantic relationship between a symbolic language and a set of concepts. A symbolic language is described by a set of basic symbols, usually an alphabet and rules that specify how symbols are combined (structured) to form more complex elements, i.e., words and phrases. One common example is natural languages. The same concept is represented differently in different languages. The information of the various representations is precisely the relationship with that concept. Of course, in addition to natural languages, there are also programming languages, mathematical languages, the language of images expressed in pixels, sign language, etc.

How is information created? Firstly, one must define the symbolic language, its symbols and its *syntax*, that is, the set of rules governing the combinations of symbols required to form complex entities, such as words and phrases. The definition does not need to be rigorous and can emerge experientially as a practical result. Nevertheless, it requires two relationships, as shown in Fig. 3.1.

One codifies *concepts* and provides their *representation* through symbols.

The other provides the *meaning* or *semantics* of the language. It decodes a symbolic representation by associating it with concepts. It should be noted that this relationship must be compositional: the meaning of a phrase is determined by the meanings of its constituent words and its structure. Therefore, the meaning of the phrase "I give the book to George" arises by composing the meanings of "I," "give," "the book," and "to George," taking into account the grammatical analysis of the structure that identifies the subject, the verb, and the two objects.

For the time being, I will not try to explain what the mind is and exactly what concepts are. It is something I consider as given—just as energy and time are considered as given, albeit recondite, in physics.

Now I would like to stress certain characteristic properties of information, and explain that, as an entity, it is different from physical entities.

- Unlike the basic entity of the physical world, which is matter/energy, information is an abstract relationship defined independently of the space-time. All our concepts and knowledge of the world are space- and time-independent, for instance, a mathematical function computed by a program. Of course, the execution time of the program depends on how fast the computer is—but that is irrelevant to the nature of the computation itself.
- One characteristic attribute of information that it is intangible, i.e., it is not subject to the laws of physics. While it needs a medium (matter or energy) to be represented, such as sound, image, or electrical waves, it is independent of the medium used and its properties. For example, when I send an e-mail, it can be read by the recipient and transmitted orally to a third party, who can place it within a text, etc. The information is the content of the message. What matters in this process of transmitting information is for the content to remain unaltered, whatever the encoding and mode of transmission.

Years ago, when I gave a lecture to gifted children in India, a young boy raised his hand and asked me: "Sir, does information weigh anything?". I asked him, "What do you think, if you erase your computer's memory, will its weight change?". He correctly answered "no." This also reminds me of a joke where a father and his son are eating Swiss cheese, the one full of large holes. The boy asks, curiously, "Dad, the cheese we eat goes to our stomach, but can you tell me what happens to the holes?". That is precisely what *information* is: *it is the structure, the organization of a medium to which we can ascribe meaning.*

Computational phenomena are transformations of information that may result in the creation, loss, or preservation of semantic content—which is a more technical matter, so let's discuss it later. A key axiom in my view is that creating information is the exclusive prerogative of the mind. Computers cannot generate information. Algorithms are transformations that preserve information defined by programmers.

While information is intangible, it matters far more than tangible goods in modern economies. That is why we speak of an intangible or a knowledge-based economy.

It is the information content of our brain that determines to a large extent who we are. Imagine that, for a moment, your memory was erased. Externally, you would be the same person, with the same physical attributes, but essentially you would no longer be "you."

In conclusion, I must point out that the concept of information as I present it is the one defined by Alan Turing, and it is foundational for informatics. It must not be confused with another concept of quantitative information, what we usually call "syntactic information," which concerns the minimum quantity of memory or any medium necessary to represent or transmit information.

I refer to Shannon's theory [4] and other similar theories that determine a measure of the information content of a structure, irrespective of what it means. These theories are useful for the information economy, i.e., to assess how much the storage (memory) or transmission (channel bandwidth) of information costs in terms of natural resources. In contrast, it is not at all useful in helping us understand what computation is.

Thus, if in the sequence of symbols "good morning" I change the order of the letters, I obtain sequences with the same syntactic information, regardless of what they mean. It is like saying that a kilogram of cotton and a kilogram of gold are equivalent because they weigh the same!

Unfortunately, many authors explain the concept of information in this way as a physical quantity and do not care about its correlation with languages, and computing. They thus adhere to the "brave" tradition of those who turn a blind eye and, when unable to describe something according to their favorite theory, prefer to ignore it rather than to question the theory's suitability.

3.3 About Computing

3.3.1 Algorithms—Conventional Computers

An algorithm is a process or set of rules to solve a problem, such as baking a cake, estimating the temperature of the planet or buying a book online. It is convenient to think of an algorithm as a procedure that allows for the calculation of mathematical functions. An algorithm that calculates a function f is a sort of "recipe" that says, for a given value of x, how one can calculate the corresponding value $f(x)$. Of course, computers run algorithms "blindly" and apply symbol transformations without "understanding" their meaning.

For technical reasons, the representation of information in computers is binary, i.e., the words of the computer language are written as sequences of two symbols commonly represented as "0" and "1".

Figure 3.2 shows the execution of an algorithm that calculates the "sum of two integers" function for the values 5 and 7. The algorithm starts with the binary representation of 5 and 7. It then calculates the result in successive steps, which once decoded, gives us the number 12. The arrows show how the algorithm combines the respective powers of two of the binary representations of the numbers 5 and 7 to provide us with the binary representation of the number 12. We usually write algorithms in programming languages. A program that describes an algorithm is a sequence of instructions that we can run on a computer.

I will not elaborate on my explanations concerning algorithms, but I would like to comment on two fundamental principles of the theory of computation.

The first principle follows from theorems proven in 1931 by Austrian mathematician Kurt Gödel, which show that, among all mathematical functions, there are

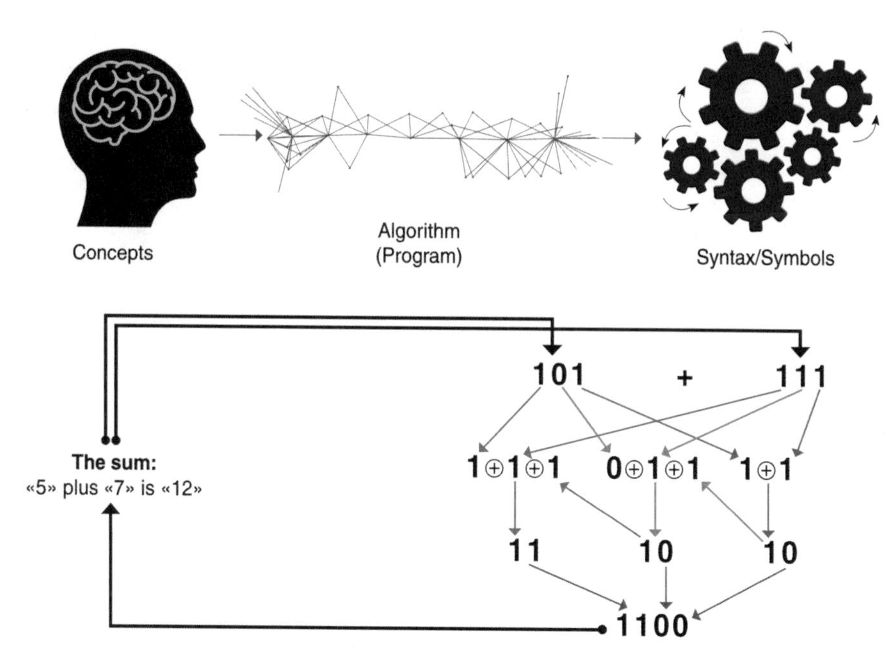

Fig. 3.2 The concept of algorithm as a symbol transformation process

functions that are not algorithmically computable. In fact, the set of non-computable functions is non-enumerable, i.e., far larger than that of computable functions, which is enumerable. This result sent shockwaves through the world of mathematics, as it clearly limited our ability to accurately calculate a huge number of functions and to solve mathematical problems. Unfortunately, many non-computable functions are very useful, such as those that allow us to check whether the programs we write are correct. For example, the veracity of the phrase "a program will terminate" or "the computer memory is sufficient to run a program" is not computable.

Gödel's results have important practical implications that will be discussed in what follows. Their immediate consequence is that we cannot use computers to check the correctness of the software we write. This theoretical limitation constitutes a type of "uncertainty principle" for informatics. Some non-computable functions can be approximated and, of course, as the accuracy of computation increases, so does the cost of the calculation.

The second principle stipulates that there is a cost for running an algorithm characterized by its *computational complexity*. The latter is the quantity of resources (memory and time) needed to execute the algorithm. Of course, the same problem may be solved using algorithms with a different complexity, but their complexities cannot fall below a certain threshold. In other words, just as there is friction in nature, meaning that when we produce work, we inevitably have energy losses, we must also consume resources when running programs. The worst case of complexity is the one where the cost of computation in terms of time or memory is an exponential function of the size of the initial data. However, even algorithms with relatively low complexity can be practically incomputable.

3.3.2 Neural Networks

Here I must point out that there are other computational models that are completely different from the conventional ones used by our computers. These models mimic natural computation processes, and there is no clear separation between the algorithm/program and the computer running it. Such models include those of natural computing, which we will discuss in what follows and which leads to different kinds of computers, such as analog, quantum, and neural networks. The latter adhere to a computational model that mimics the neural networks of our brain.

While the principles of neural networks have been known to us since the mid-1940s, advances in research have enabled them to be applied with marked success to problems of artificial learning over the last two decades, giving a huge boost to the development of artificial intelligence and its applications.

Neural networks allow for the effective solution of problems whose algorithmic solution is hampered by an insurmountable complexity barrier. Let us take a simple example by assuming that I want to build a system where I input images of cats or dogs, and the system correctly identifies the animal being depicted.

The traditional algorithmic approach starts with an analysis of models that characterize each animal's form with a set of patterns, such as the shape of the animal's head, the position, and shape of its eyes, ears, and nose. Based on this analysis, I would have to write an algorithm that analyzes images, identifying characteristic patterns, and deciding whether the image is a dog or a cat. This *model-based* algorithmic approach is very difficult to implement when analyzing composite images and is computationally very costly.

Machine learning takes a completely different, empirical *data-based* approach that does not require model analysis and programming. A neural network is a system that, for each information being input, generates a corresponding response. The network learns to distinguish cats from dogs through an experimental process, just like a child. In other words, we "show" the system a large number of related images and progressively configure its parameters so that it provides the correct response. The "training" process of the system is adaptive: the percentage of incorrect responses decreases as the size of the training dataset increases.

Of course, a child learns with just a few examples, while training neural systems requires a large volume of data. However, their advantage over traditional systems running algorithms (programs) is that, after potentially time-consuming "training," neural systems compute their response almost instantly. As a result, companies such as Nvidia and Waymo have developed neural systems that are capable, after intensive "training," of driving cars with relatively small but still non-negligible error rates.

3.4 Knowledge and Its Validity

3.4.1 What Is Knowledge?

Knowledge is information which, incorporated into the right network of semantic relationships, can be used either to understand a situation or to act in order to achieve a goal.

This definition considers knowledge to be useful information, and it should therefore have a certain degree of truthfulness and validity. The phrase "the sun revolves around the earth" is information, but it is not knowledge, even though it was considered to be knowledge in the past. A recipe for baking a pie or the solution to a mathematical problem is knowledge insofar as it enables us to achieve corresponding goals.

It is important to understand that knowledge, as we have defined it, has a general and dual character (Fig. 3.3).

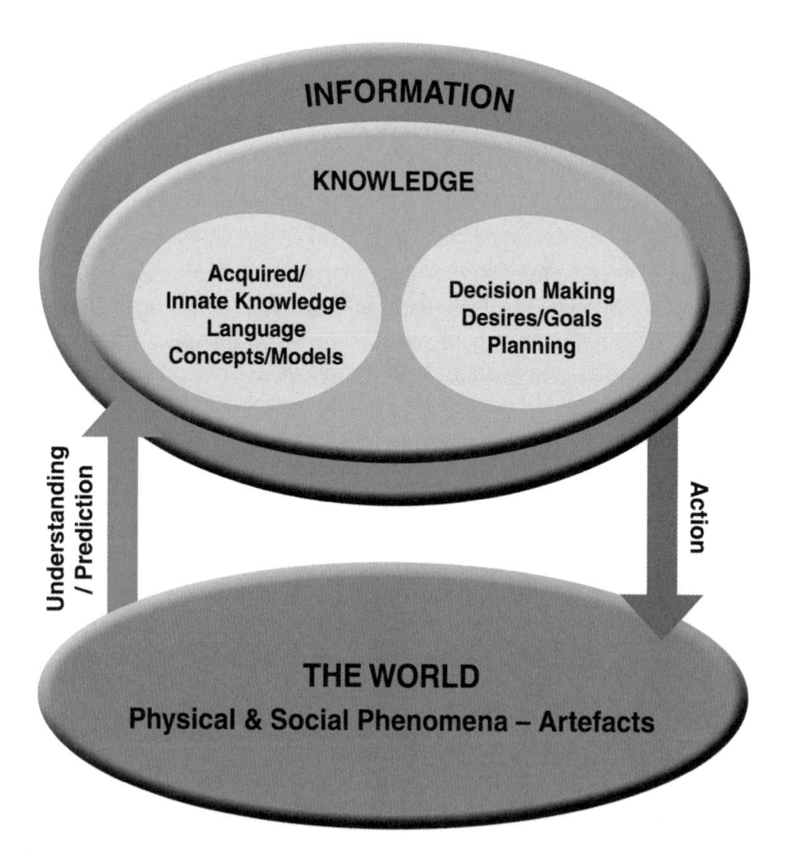

Fig. 3.3 Knowledge as information and its dual use

- On the one hand, it allows us to understand situations, i.e., what is happening around us, which includes the physical world, societies, and their artifacts. It can be a simple item of information that holds true at a given moment in time, such as a measurement (my maximum blood pressure is 14), the statistical result of a survey poll (45% positive opinion) or experimental knowledge (the speed of sound is 343 m/s). This type of knowledge includes both scientific and shared empirical knowledge.
- On the other hand, it allows us to achieve a goal. In this case, knowledge is linked to meeting needs and entails action to change the state of the agent or of its external environment. This type of knowledge includes technical knowledge, such as that of an engineer (in order to design a house) or a programmer (to write a program). It also includes implicit empirical knowledge that allows us, for example, to walk or to speak.

Below, I propose a classification of the different types of knowledge according to their validity and applicability. I also explain that there are different ways of developing knowledge that also determine how it is transmitted and used.

At this point, I would like to emphasize a fundamental tenet of this book. The world can only be understood as the combination of phenomena of the material world and its mental representations. This dualist vision of the world distinguishes between two basic entities: matter/energy and information, corresponding respectively to what can be observed in the material world (phenomena) and what can be thought (noumena). This is a very important distinction so as not to confuse, as many often do, "the map with the country," i.e., the world with its models. In this context, the human mind plays a key role. It is a "supercomputer" that either instantly creates knowledge by explaining the world or takes action by applying knowledge in order to achieve goals.

The above position differs from others, which consider matter/energy as the only fundamental entity and, therefore, mental phenomena as emergent properties (epiphenomena). I hope to demonstrate in the following chapters that such positions have inherent weaknesses. They are unable, taking a bottom-up approach, starting from the particle world, to explain mental phenomena, such as language and consciousness, as emergent properties of the brain.

Similar problems also arise with positions that prioritize an intangible entity such as information, taking a top-down approach.

3.4.2 *Types of Knowledge and Their Validity*

The issue of the validity of knowledge has been much debated and is very timely because knowledge is produced and used not only by people but also by computers. Many believe that there is a wall between scientific and non-scientific knowledge. However, I will show that subtler distinctions may exist.

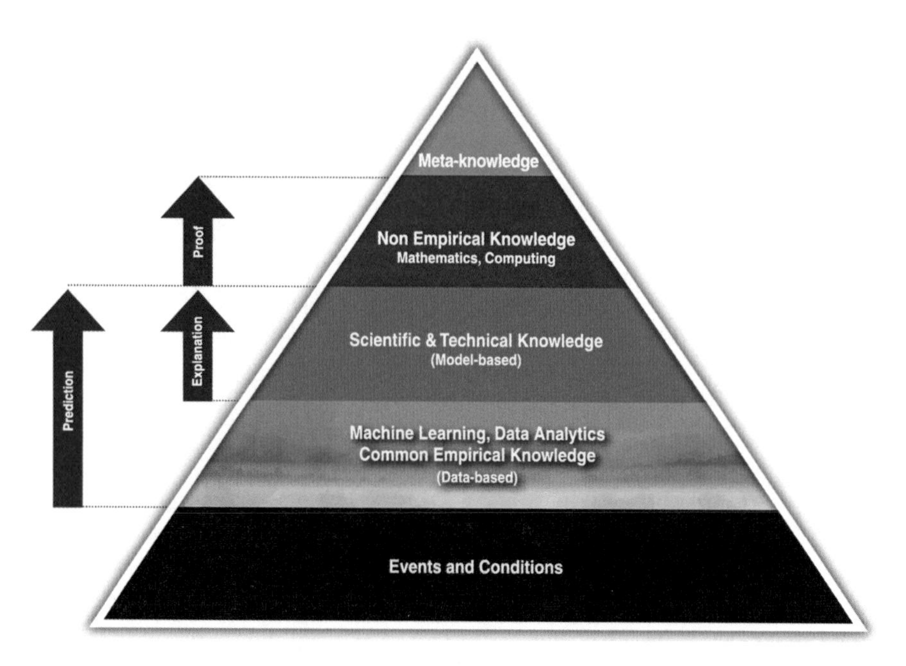

Fig. 3.4 The knowledge pyramid—classification of types of knowledge

Figure 3.4 proposes a hierarchical classification of the different types of knowledge produced by humans or machines.

One significant difference is that between *empirical* and *non-empirical knowledge.*

Empirical knowledge is acquired and developed through our senses based on observation and experimentation. It includes all the knowledge we have gained automatically through our experiences or after informed mental processes.

Non-empirical knowledge, which is also called *a priori knowledge*, is the product of mental processes that are not directly related to our experiences. Its validity depends only on the consistency of our reasoning. It expresses relationships that remain unaltered and independent of space-time, where physical phenomena occur. It includes, among other things, mathematics and logic, as well as theories of computation. Concepts such as straight line, plane, and unit, constitute a priori knowledge. The Pythagorean theorem or Gödel's theorems are "eternal" truths, propositions whose validity depends only on the axioms of Euclidean geometry and arithmetic, respectively.

A higher kind of empirical knowledge is that based on mathematical models. This includes *scientific knowledge* for a reasoned understanding of the world and *technical knowledge* for the safe and effective achievement of goals. These types of knowledge are discussed in detail below. They differ from common empirical knowledge in that they constitute a generalization of observations embodied as mathematical models. This allows for an *explanation* of the relationships (laws) between variables representing the observed quantities, for example, that acceleration is proportional to force or that the drag force of a moving object is proportional to the square of its speed. The existence of such relationships, which we call *physical laws*, also allows for what we call *predictability*. Once there is a law characterizing a phenomenon, it is possible to predict the impact of the change of some variables (causes) on its overall dynamics.

The simplest type of empirical knowledge is *facts* and *situations* in a given place and time. For example, the phrase "the temperature in Athens today is 25 degrees" describes a situation, while the phrase "the Battle of Waterloo took place on Sunday, 18 June 1815" is a fact. This type of knowledge is of limited applicability but is necessary to understand the world.

Most of our knowledge is *common empirical knowledge*. Much of our intelligence relies on it. It is the result of a generalization of empirical knowledge, which may take place consciously or subconsciously. Its validity is simply based on its practical use.

It is thanks to this kind of knowledge that we can walk, speak, play instruments, dance, etc. It also includes all skills and knowledge based on common sense, such as implicit relationships we consider "obvious." We therefore know that parents are older than their children or understand the dynamics of the kinetic states of objects, even if we do not know the laws of physics (see Sect. 6.1.2).

I include the data-based knowledge generated by computers, such as that produced by neural networks, in common empirical knowledge. It is clearly empirical and of the same nature as the common empirical knowledge of humans. It is knowledge that allows for predictability, as a neural net computes a relationship between inputs (causes) and outputs (effects). However, it does not have the validity of scientific knowledge, insofar as we cannot generally describe this relationship using mathematical models, as I explain below.

Finally, knowledge generated through data analytics techniques can be either empirical or scientific when they are based on statistics theory.

At the top of the pyramid is *meta-knowledge*, which some call "wisdom." It is knowledge on managing any form of acquired knowledge. It allows us to combine knowledge of all kinds. It includes problem-solving, design, and decision-making methods. It also includes non-formalized knowledge used in professional skills.

The above classification demonstrates that, depending on how it is developed, knowledge has different degrees of validity and generality. It allows us to understand the differences between natural and artificial intelligence, which will be discussed extensively in the following chapters.

3.4.3 Scientific Knowledge

3.4.3.1 The Process of Developing Scientific Knowledge

Scientific knowledge helps us understand the physical world. It is mainly based on analytical thinking, which connects phenomena to the world of mental models, information, and mathematics. As I have said, its development leads to the formulation of *laws*, unalterable relationships that govern the changes observed, for example, conservation of energy, universal gravitation, gas laws, and the law of supply and demand. It follows three steps whose realization raises different kinds of problems.

1. *The first step* aims at capturing the observed reality through a model. For example, if I want to study meteorological phenomena, I will first build a model based on a number of empirical assumptions. The model may be formalized in a purely mathematical manner, for example, a system of equations, or in a less rigorous and ad hoc manner. It expresses relationships between quantities such as pressure, temperature, direction, and speed of winds. Of course, I can create different—and not necessarily comparable—models for the same phenomenon. This is determined by what is regarded as observable and measurable.

   The difficulty of modeling a phenomenon is characterized by what is called its *epistemic complexity*, i.e., how difficult it is to find an appropriate mathematical or mental model for the phenomenon we are studying. For example, let us assume that someone is experimenting in an effort to understand a phenomenon where a quantity increases exponentially. If the experimenter was, for instance, the ancient mathematician Thales of Miletus, who only understood proportionality relationships, then he would not be able to explain the phenomenon.

   We know that, in order for Newton to formulate his theory, he had to develop the relevant mathematical models (differential calculus). Without these models, the discovery of the theory of universal gravitation would have been impossible.

   By this, I simply mean to say that the scientific laws we discover depend on the models at our disposal. There are extremely complex phenomena for which we cannot find suitable models, thus limiting the possibilities of advancement of scientific knowledge. For example, understanding human behavior is faced with a practically insurmountable "wall" of epistemic complexity.

   Note that sometimes the inability to study complex phenomena is not so much due to a lack of theory, but to our inability to formulate a model that faithfully represents the phenomenon by combining a large number of parameters and facts.

   For example, let us consider a system with two teams of robots playing football. I could theoretically predict the outcome of the game if I could build a faithful model of the system: accounting for each player (as an electro-mechanical system with hardware and software), for the objects in its environment and how they interact. The explosive difficulty of the problem does not lie in studying new phenomena or discovering new laws. It concerns the faithful description of all the extremely complex relationships that characterize the dynamics of the phenomenon.
2. *The second step* in the discovery of scientific knowledge concerns the analysis of the model, carried out by experts usually assisted by computers. The analysis will reveal relationships between the observed quantities of the model, either in the form of explicit mathematical relationships, for example, cause-and-effect relationships, or in the form of graphical representations of the phenomenon, usually

generated by computers. Thus, an accurate weather prediction achieved by resolving a meteorological model depends both on the model's faithfulness and on the accuracy of the resolution method.

Thanks to the use of computers, we can analyze complex models and push the boundaries of knowledge as far as possible. At this point, I must recall the limitations concerning *computational complexity* (see Sect. 3.3.1). If a precise resolution of the model is not feasible, then we may be able to find approximate solutions that are satisfactory in practice. The higher the accuracy of resolution of a model, the better the predictability of the phenomenon.

I noted above that there might be more than one model for the same phenomenon, representing it in different ways or in different degrees of detail. Detailed models primarily allow for better predictability, but their analysis may involve increased computational complexity.

3. *The third step* in the discovery of scientific knowledge is to verify the faithfulness of the model. This means that if a relationship holds true for the model, for example, a prediction concerning tomorrow's weather, this will also be confirmed, with a certain tolerable error margin, by observing the phenomenon under study. Of course, confirmation requires comparison of a sufficiently large number of experimental data and an investigation demonstrating that the data considered "sufficiently" cover a representative set of states of the phenomenon.

   This raises certain methodological problems, as physical phenomena have, in theory, an infinite number of states that cannot be investigated. Error is therefore logically possible, however large the number of experimental confirmations.

   This context gives rise to the problem of exploring the entire spectrum of states in the best possible way and concerns the *controllability* of the phenomenon under study. It is the possibility of bringing it into predefined initial states, which is not feasible, for example, for numerous geophysical, astrophysical, economic, and social phenomena. The distinction between controllable and uncontrollable phenomena creates a dividing line as to the validity of scientific truth. In classical physics, phenomena are largely controllable. We can reproduce the phenomenon starting with the same initial conditions and make it repeatable, which confirms that our model's predictions are not random.

   There are many phenomena that we cannot reproduce for various reasons. Some, such as social phenomena, are not repeatable by their very nature. Some phenomena could be reproduced with some probability, while others are difficult to bring to exactly the same initial conditions, for example, economic systems. For a scientific field, reduced controllability implies decreased validity of knowledge. Nonetheless, this limitation should not imply any disparagement of the field.

In summary, applying the scientific approach and producing scientific knowledge involve three steps and corresponding complexities (see Fig. 3.5).

1. Epistemic complexity, which depends on our ability to grasp a phenomenon using existing mathematical models and modeling tools.

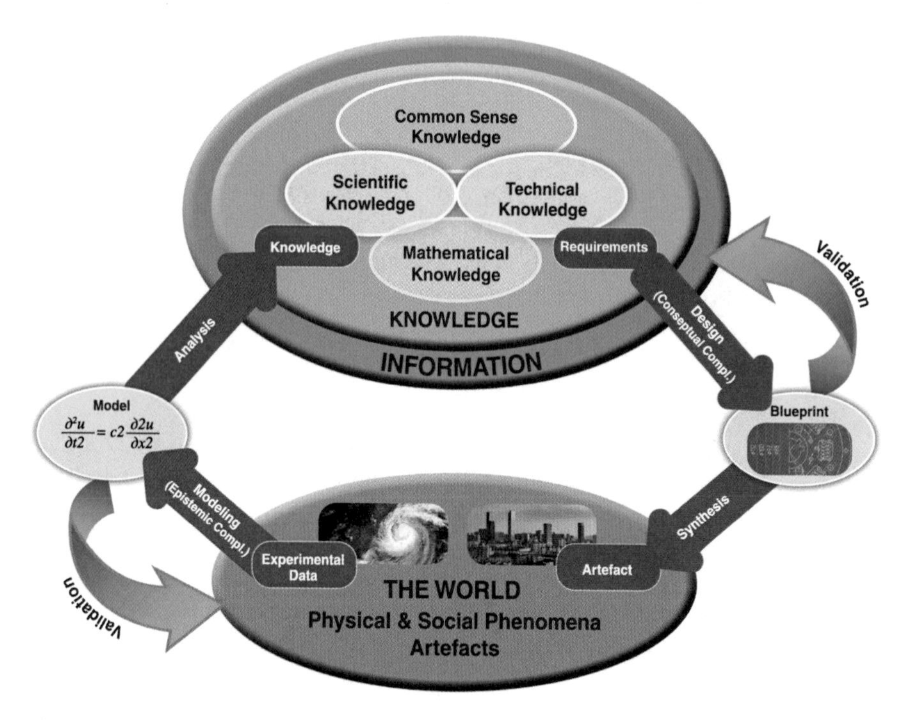

Fig. 3.5 Development of scientific and application of technical knowledge

2. Model resolution complexity, which is mainly computational and depends, in turn, on the complexity of models and the effectiveness of resolution techniques. The latter is determined by computer performance and the availability of suitable model resolution algorithms.
3. Validation complexity, which depends on the degree of controllability and the dynamics of the phenomenon under study.

I would like to stress that the proposed methodology is more general than others limiting the development of scientific knowledge to cases where phenomena are controllable and can be experimentally reproduced. I believe such a limitation makes no sense. On the one hand, the degree of controllability of phenomena may vary. On the other hand, it denies entire fields of knowledge, such as social sciences, scientific status.

3.4.3.2 On the Nature of Scientific Knowledge

Having explained the process of developing scientific knowledge, I would like to highlight some of its features.

Our knowledge of physical, economic, and social phenomena is described by laws which express relationships between observed quantities in space-time. These "laws" are not of the same nature as the laws of a legislative system, whose violation entails criminal liability. Nature is not responsible if it does not comply with what we consider to be its laws.

We have seen that the development of scientific knowledge depends on the ability to find appropriate models and concepts provided by the language of mathematics and logic. Therefore, the laws of phenomena are "invented" by people rather than "discovered." Some people wish to believe that these laws predate human thought. This is an ontological issue, and I will not discuss it further.

Laws are created in the mind of the observer as a *generalization of relationships*. In other words, by ascertaining that a mathematical relationship is validated between observed quantities for a large number of experiments, we declare this relationship a law. Such a leap in logic involves arbitrariness, as no validation can be made regarding the infinity of possible combinations of values. For a set of values, the observed relationship could be falsified—and this was indeed the case when the theory of relativity demonstrated that Newton's simple and elegant theory of universal gravitation is not valid for high velocities.

This generalization is called *physical induction* and is different from *mathematical induction*, which is based on axioms and produces mathematical knowledge (see Sect. 4.1.1.2 Formalized Language—Theory).

Scientific knowledge therefore states, with some degree of certainty, what can happen "here and now." No matter how many observations satisfy the inductive relationship, there is no guarantee that the next observation will not violate it, thus definitively canceling its validity.

The renowned mathematician and philosopher Bertrand Russell derides the confidence of physicists regarding the power of scientific truth with the story of the "inductivist turkey" [5]. It is about a turkey on an American farm that decided to construct a scientific view of the world in which it lives. Thus, on its very first day on the farm, it observes that it is fed at 9:00 in the morning. Of course, it did not rush to draw conclusions. It waited for the next day and found that it was at 9:00 in the morning, and then every day of the week, rain or shine. Therefore, ultimately satisfied and considering that it had a fairly large number of confirmed observations, it inductively concluded that "I am fed every day at 9:00 in the morning"—that is, until Christmas Eve, when, instead of being fed, it had its throat cut.

Unfortunately, the relativity of scientific knowledge is yet to be sufficiently emphasized or understood, even by famous scientists. What is one to think of the assertion of the famous physicist Stephen Hawking who believed that "the universe and the "Big Bang" was an inevitable consequence of the laws of physics" [6]?

3.4.4 Technical Knowledge

Technical knowledge uses synthetic thinking. It is used to build artifacts that meet given techno-economic goals to satisfy human needs.

There are many examples of complex systems challenging our technical knowledge. One, for instance, concerns autonomous transport systems and self-driving cars, in particular, which we hope to see in a few decades. Another "distant" system is the "Next Generation Air Transportation System" which envisions the complete automation of air traffic control [7].

The development of technical knowledge for building artifacts consists of three steps, the realization of which raises different kinds of problems (Fig. 3.5).

1. *The first step* is to collect, understand, and formalize the needs leading to the design of the artifact. Needs are initially expressed as a set of requirements in natural language. After having been analyzed, requirements are organized by experts into technical specifications, which form part of the contractual agreement between the system developer and the stakeholder (client). This text, many hundreds of pages long in cases of complex systems, must be checked according to two criteria: (a) logical consistency, i.e., that it does not contain contradictory statements; and (b) completeness, i.e., that the specifications adequately cover all the required operational needs. This check is carried out by specialized engineers aided by computers. However, its automation is hampered by difficulties inherent to formalizing and analyzing natural languages.

   For some application areas from the realm of physics, the formalization of needs may be relatively easy, for example, for an electrical circuit, insofar as its technical characteristics can be formulated using existing theories. For other applications, for example, flight control systems, the problem is harder as the precise formulation of their properties requires mathematical languages that are complex and difficult to use.

   Another difficulty is the comprehension of needs of users involving criteria that cannot be objectively defined. For example, functional and esthetic requirements that are difficult to formalize may be added to the technical specifications of a house.

   The difficulty in standardizing needs is characterized by what we usually call *conceptual complexity*, i.e., the difficulty of formulating and structuring descriptions in a given language. It is obvious that the formulation of the technical requirements of a bridge is far simpler than those of a self-driving car. In the first case, we can amply rely on concepts from mechanical engineering, while in the second we must address, in addition to technical issues, other badly formalizable issues such as traffic regulations and even legal aspects, for example, the driver's liability.

   Once needs have been definitively formalized, the experts proceed with designing the artifact by following methodologies that vary depending on the application sector. Methodologies determine the organization of design work carried out by engineering teams assisted by computer-aided design tools.

   The aim of the overall process is to design an artifact in the form of a model that meets the initial requirements. Such models include the blueprints of a civil engineer, the wiring of an electrical system, or the software of a computing system. Since the process cannot be automated, there is always a risk that the result does not meet expectations, which we will discuss later.

2. *The second step* in the development of technical knowledge for building artifacts, concerns using the design model to synthesize a construction model that sets out a manufacturing process. The synthesis matches elements of the design model with functionally equivalent material components, taking into account their physical properties such as their strength and resilience. We can choose different qualities of materials for the same design, provided, of course, we meet the original technical requirements. This is why the synthesis also involves design space exploration aiming at optimizing construction costs. Based on the construction model, we can determine the techno-economic characteristics of the artifact, such as construction costs, safety, and performance.

   Depending on the technical area, the synthesis of the construction model can be largely automated. Thanks to computers, we can build complex aircraft, trains, factories, enormous bridges, and buildings. However, the most complex artifacts at present are undoubtedly computerized systems, whose design and construction would be impossible without the help of computers.
3. *The third step* in the development of technical knowledge for building artifacts, concerns validating that the construction model meets the original requirements. The approach is similar to that for validating scientific truth, except that it is usually not carried out on the artifact but by simulating the construction model. Once again, the problem of covering and investigating a large number of states arises. Fortunately, the use of simulation techniques allows for better controllability. There are various validation techniques, the most commonly used of which is *testing*, which involves an experimental analysis of the behavior of the artifact: application of scenarios and comparison of behavior with the one envisaged in the initial requirements.

We have described the general principles for the development of technical knowledge in order to build artifacts. Three types of obstacles and corresponding complexities also arise in this case (see Fig. 3.5):

1. Conceptual complexity, which expresses the inherent difficulty in describing the technical specifications for the artifact to be built.
2. Complexity of synthesis of the construction model, which is mainly the complexity of computing optimal construction solutions.
3. Complexity of validation, which is mainly the complexity of exploration and coverage of a sufficient number of states confirming that requirements have been met.

We will demonstrate in the next chapters that the difficulties vary depending on the technical domain. For some sectors, there is no need for validation at all, while for others, such as computer systems, validation costs account for a significant part of the development cost, for example, over 50% for large systems.

I would also like to point out that the development of a relatively complex information system involves significant risks of partial or complete failure. If we measure complexity in lines of code, the possibility of failure in the development of a system with over one million lines of code, such as an operating system, is approximately 30%, while the probability of partial failure is over 50%. Interested readers can find numerous relevant sources online, such as [8].

3.4.5 A Methodological Clarification

To conclude this chapter, I would like to point out that I adhere to a definition of knowledge that is broader than that found in philosophy books and dictionaries, where it is usually equated with scientific knowledge. It is worth noting that in English the field of philosophy that deals with the study of knowledge is called "epistemology," a term that relates to knowledge for understanding the world, particularly scientific knowledge.

The *Oxford Dictionary* provides the following definition for "science" [9]: *The intellectual and practical activity encompassing the systematic study of the structure and behaviour of the physical and natural world through observation and experiment.*

Other dictionaries provide similar definitions. Their common feature is that science is interested in discovering facts and laws that govern the world. Thus, while physics and biology are considered sciences under that definition, mathematics and informatics are not pure sciences. This is because, as I explained, even though a large portion of mathematics was developed in order to study physical phenomena, mathematical truths are completely independent of these phenomena. I would say that the same holds true for the principles of informatics. Furthermore, this definition does not include anything related to the application of scientific knowledge—for example, engineering or medicine are not pure sciences.

Note that the acceptance of a narrow definition of knowledge that links it to science alone has occasionally led to sterile discussions.

Another problem with the current definitions of science is that they overstress the importance of the experimental method. Thus, some people doubt that economics or psychology are sciences. Others consider that there are impermeable limits between science and non-science.

In order to avoid pointless disputes, I believe that knowledge has degrees of validity, which depend on whether it is amenable to experimental confirmation and the degree to which it fits reality. Observation of astronomical phenomena offers limited possibility of repeatability. However, such restrictions challenge us to devise methods that will push the boundaries of knowledge further and improve its validity.

I believe that we should get rid of narrow definitions that call into question the uniform nature of knowledge. We must focus on the concept of knowledge in all its forms, as it is connected to computing, that is, information that is useful for solving problems, a concept that embraces scientific knowledge, technical knowledge, mathematics, and general empirical knowledge, which becomes particularly important through artificial intelligence.

As for those who overstress the importance of scientific knowledge and its "superiority" over technical knowledge, I would remind them that their development has been concurrent and complementary over the centuries. The ancient Greeks clearly distinguished "*episteme*"—science from "*techne*"—art, and the latter had a broader meaning than it does today, initially referred to skillfulness accompanied by knowledge in order to perform a work or profession. Great mathematicians and physicists such as Archimedes and Heron of Alexandria were famed engineers.

Furthermore, the advancement of mathematics and physics during the Renaissance was largely carried out by engineers such as Galileo and Leonardo da Vinci.

Today more than ever, scientific knowledge and technical knowledge are intertwined in a virtuous cycle of mutual interaction and growth. Scientists build complex experiments to study natural phenomena, while engineers need increasingly sophisticated theories in order to build systems.

To sum up, I would like to stress that knowledge should not be compartmentalized. Knowledge has varying degrees of validity and multiple ways of development, especially thanks to computers. Advances in medicine and biology, physics and chemistry have been astounding. Even mathematics and logic, an exemplary form of abstract knowledge, acquire an experimental dimension thanks to computers, which are used to prove theorems and establish results.

Modeling and simulation techniques enable the study of complex physical phenomena, such as the flow of lava or the behavior of complex information systems. Models might not be directly related to observation and experimentation; they may be entirely ad hoc or may combine theoretical and empirical results. What truly matters is whether the results are consistent with the observations of the phenomena being simulated, and whether they allow them to be understood and predicted.

Contrary to the prevailing view in physical sciences, the development of knowledge does not always start with the observation of phenomena and experimentation. This holds true for mathematical knowledge, but sometimes for scientific knowledge as well. The theory of relativity indeed began with observations but was developed in a theoretical framework based on thought experiments that were validated by observation at a later stage. The theory of computation is based on a priori mathematical knowledge. A doctor uses complex technology to diagnose a disease, making assumptions that could explain the symptoms. This is followed by a trial-and-error process in order to reach the most well-founded diagnosis. Failure to comply with the "classic" methodology of developing knowledge in the physical sciences does not logically imply invalidity.

As explained in the Introduction, instead of the term "epistemology," I prefer the term "gnoseology," which also exists but is less prevalent. However, it is more appropriate as it avoids the misunderstanding that knowledge is limited to comprehending phenomena. I do, of course, employ the term "epistemology" when referring to the study of scientific knowledge.

In conclusion, I would like to once again emphasize that knowledge today is more important and strategic than material goods. It allows for dominance over the physical world and response to global challenges, provided, of course, we possess the meta-knowledge to manage it properly.

References

1. https://en.wikipedia.org/wiki/Edsger_W._Dijkstra
2. *The Blind Watchmaker*, Richard Dawkins, 2015.
3. "Brain could exist outside body – Stephen Hawking", *The Guardian*, Sept 23, 2013.
4. https://en.wikipedia.org/wiki/Information_theory
5. https://mashimo.wordpress.com/2013/03/12/bertrand-russells-inductivist-turkey/
6. https://www.reuters.com/article/us-britain-hawking-idUSTRE6811FN20100902 (accessed on 08/01/22)
7. https://en.wikipedia.org/wiki/Next_Generation_Air_Transportation_System#History
8. https://www.standishgroup.com/sample_research_files/CHAOSReport2015-Final.pdf
9. https://www.lexico.com/definition/science

Chapter 4
The Development and Application of Knowledge

4.1 The Development of Knowledge—Principles and Limitations

Although the development of knowledge is governed by general principles which I explained above, it is useful to distinguish between *fields of knowledge* according to their subject matter, which is defined by their fundamental concepts and relationships and the body of knowledge developed on their basis.

Mathematics and logic are the most fundamental fields of knowledge as they provide models and theoretical tools to represent the world and its phenomena. They are necessary for every other field of knowledge, insofar as its specialized knowledge can, of course, be formalized and studied as mathematical relations.

Following mathematics, I consider key fields of knowledge to be physics, informatics, and biology, as they have separate and non-overlapping subject matters.

Physics is concerned with the study of matter/energy phenomena in space-time.

Informatics is concerned with the transformation of information—what we can compute and how to go about computing it. It is not a field of mathematics, but it is based on mathematical models. Additionally, it has an important technical component related to building computers.

Biology is concerned with biological phenomena involving intertwined physicochemical and computational processes.

The other fields of knowledge can be considered composite since, apart from specialized knowledge, they also use knowledge from these key fields. For example, economics is a composite field of knowledge because it deals with relationships between natural and monetary resources managed by people. The main problem is how to best use resources in accordance with specific goals in order to meet individual and social needs. The study of economic systems combines domain-specific knowledge and more general knowledge from various fields such as psychology, sociology, automatic control, and systems theory.

J. Sifakis, *Understanding and Changing the World*,
https://doi.org/10.1007/978-981-19-1932-9_4

All fields of knowledge, regardless of their subject matter, have a common approach characterized by the following three principles:

1. *Modeling*, i.e., using models that are representations of phenomena and systems expressed through natural or artificial languages.
2. *Stratification* of modeled reality through an appropriate abstraction hierarchy highlighting key properties to be studied for each modeling layer.
3. *Modularity* of the models used in each layer to break down the complexity of the modeled reality and understand it as a synthesis of components.

4.1.1 Modeling—The Role of Language

Modeling is the process of representing a phenomenon or a system under study. It is done using either a natural language or an artificial language, usually based on mathematics, such as systems of equations, algorithms, or logic. As a rule, natural languages allow for freedom of expression, as they are rich in concepts and structures. Conversely, artificial languages are built with a limited number of well-defined basic concepts and structures. Their advantage over natural languages is the analytical ability of mathematics. Generally, expressiveness and analyzability are inversely proportional: the more expressive a language is, the lower its analyzability.

4.1.1.1 Natural Languages

Ludwig Wittgenstein considers that "the limits of my language are the limits of my world," i.e., our language sets the limits of our understanding of the world [1].

Concepts and language, as they took shape empirically, are the first models of the material and mental world. As for how they emerged through evolution and became vehicles for cognition is another story, about which we can but speculate. It seems that humans grasped physical numbers and the basic operations between them at a relatively early age—as seen, for example, in the Sumerian astronomical tables. Humans also tried to use myths to explain natural phenomena or to codify certain ethical rules quite early. First, we created a kind of "technology" and "know-how," enabling us to solve a number of problems empirically and systematically. For example, we were able to build wheels, light fires, measure quantities and areas, cultivate land, and reproduce plants.

The first attempt to rationally interpret reality with what is called *theory* took place in Greece. Pre-Socratic philosophers attempted the reduction of reality into a few constitutive elements, seeking explanations based on laws. We thus saw the advent of mathematics and the study of the properties of numbers with the Pythagoreans, Democritus' atomic theory, Zeno's paradoxes of the relationship between the continuous and the discrete, Plato's ideas, and Aristotle's ingenious taxonomies. These were followed by major steps in logic by the Sophists, the axiomatization of geometry, and the development of astronomy, which led to the heliocentric model and the proof of the Earth's sphericity.

The Renaissance saw the second major wave in the further development and formalization of knowledge. Progress in mathematics was closely linked with the development of mechanics, optics, and chemistry.

It is awe-inspiring that the human mind discovered simple mathematical and geometric concepts by making powerful abstractions. We thus invented the concept of a straight line, a circle, or a right angle. The physical world is rugged, it has hardly any right corners and very few perfectly circular shapes. The human mind invented the main abstract concepts underlying geometry, arithmetic, and continuous mathematics, which have enabled us, to some extent, to grasp the essence of physical phenomena, and to achieve predictability.

The long trajectory of the intellect toward understanding the world is characterized by abstraction. It starts with the creation of names, where the mind carries out generalization—it gives the name "stone" to a set of objects in which it discerns common characteristics, despite differences in shape, color, texture, etc. This is the foremost act for the mental interpretation of the world.

Generalization creates relationships of equivalence. This allowed for the creation of categories of objects, leading to a first hierarchy of concepts. The emergence of adjectives or adjectival modifiers plays an important role here. In the ancient Greek language, one category of fruit was *karya*, with a hard shell, and another was *mela* (with a soft rind). There are various kinds of *karya* with modifiers such as Pontic (hazelnuts) or Persian (peaches), and various kinds of *mela*, such as Cydonian (quinces) or Damascene (plums).

The advent of verbs and adverbs is the first sign of understanding change and time. The basis for the invention of verbs is the construction of relationships linking words, cause-and-effect relationships, or temporal relationships. First, there were many irregular types, which abound in Homer's poems. Over the course of its evolution, language discovered laws of economy, systematizing the construction of tenses and conjugations.

Other animals also possess simple languages to specify objects, actions, and situations. However, the appearance of abstract concepts is a huge leap that makes humans stand out. Much has been written about this, particularly thanks to the analysis of the etymological provenance of concepts in old languages such as Greek or Hebrew. Humanity owes much to ancient Greek thought, which bequeathed the vast majority of philosophical, scientific, technical, and mathematical concepts.

The emergence of scientific concepts and theories is based on the power of abstraction. These include the discovery of the properties of time and space in order to understand natural phenomena. In physics, fundamental abstractions are those concerning matter and energy. By means of the "game" of successive abstractions, we come to understand the physical world through laws that express relationships between these two concepts that are characterized by measurable quantities.

Additionally, important was the use of money for transactions and, much later, understanding the economy through the capital–labor relationship.

Finally, other fundamental concepts are those of information and knowledge, the importance of which we only grasped in the twentieth century. Theories of computation allow for the mathematical study of languages.

We can view the development of scientific knowledge as an attempt to devise new languages rooted in mathematics with strong empirical foundation.

4.1.1.2 Formalized Languages—Theories

As we have said, mathematics and logic allow us to formalize knowledge in other fields. When a relationship is expressed in mathematical terms, for example, as an equation, it is strictly and unambiguously defined. On the other hand, relations expressed in natural languages may be open to multiple interpretations and are generally ambiguous.

Every mathematical theory is based on a set of *axioms* and *rules* for the concepts it formalizes. Axioms are characteristic properties of theoretical concepts and are generally inspired by empirical knowledge. We thus have the axiom that for every number n, the number $n + 1$ is different from n. Another axiom is that a proposition can be either true or false, but not both.

The rules of a theory allow us to infer theorems through axioms. One characteristic rule of propositional calculus is that of deduction: if a is true and if a implies b, then b is true. This rule is followed in the well-known example: *All men are mortal* and *Socrates is a man*, therefore *Socrates is mortal.*

Euclidean geometry is the first axiomatized theory proposed by Euclid of Alexandria in the fourth century BC. It thus demonstrated the power of formalization and paved the way for logic and abstract mathematical thinking. Only five axioms and three basic concepts are sufficient to formalize plane geometry.

One abstract concept that is very important for mathematics and their application is that of infinity and, more generally, infinite quantities. It has been proven that there are different types of infinity, and an order relation can be defined between them, similar to the order between natural numbers.

How can we prove relationships between the infinite elements of a set? This requires a different kind of logical rule: the rules of induction. These rules apply to hierarchically structured sets and allow for generalization on the basis of the following reasoning: "If a property is valid for the smallest elements of the hierarchy and its validity for layer n implies its validity for layer $n + 1$, then the property is valid for all the elements of the hierarchy."

One classic example is the induction on the set of natural numbers with the hierarchical relation $n < n + 1$, which we learn in high school. The general application of induction to infinite sets requires the definition of an appropriate hierarchical relationship, which generally depends on the type of property we wish to prove. For example, if I wish to prove that a program terminates, I need to find a function associating a positive quantity to program states and such that for all program runs, this quantity steadily decreases.

One consequence of Gödel's theorems (see Sect. 3.4.2) is that there is no algorithm that decides program termination. This result entails that discovery of induction rules cannot be automated.

Contradiction arises in a theory when a proposition and its negation are true. Contradictions violate a well-known principle of reasoning. A mathematical theory must be "consistent," i.e., it must not contain contradictions. It should also be "complete," i.e., every true proposition must be provable using axioms and inference rules.

In the early twentieth century, the German mathematician David Hilbert proposed a research program aimed at proving that mathematics is consistent and complete. Furthermore, he aimed to prove that mathematics is decidable, i.e., there is an algorithm that decides the truth or the falsity of any proposition in the system.

I have already mentioned Gödel's theorem, which states that in powerful enough consistent theories there are undecidable propositions (see Sect. 3.4.2). Gödel also showed that theories of arithmetic (integers with the operations of addition and multiplication) cannot be proven to be consistent—that is, free of contradictions—within the theory, even when they are consistent.

One way of violating the consistency of a theory is to allow self-reference, i.e., propositions that refer to themselves. The "liar" paradox, which is attributed to the philosopher Eubulides of Miletus, who lived in the fourth century BC, has been known since antiquity: "A man says that he is lying. Is it what he says true or false?". By applying the rules of logic in this case, we reach the well-known "vicious cycle." The proposition implies its negation, which implies the proposition.

This paradox remained unexplained for 2000 years, until Bertrand Russell formulated the paradox in set theory in the early twentieth century, proving that the standard model of set theory cannot prove its consistency.

One way of understanding the practical significance of these results is that any consistent theory encompassing arithmetic is "open" in the sense that there are propositions that cannot be proven or disproven in the theory.

A scientific theory condenses and formalizes scientific knowledge derived from the process described in Sect. 3.4.3. It takes its models from a mathematical theory to which special axioms are added that reflect the fundamental laws governing the phenomena we are studying.

For example, in Newton's theory, the mathematical theory is differential calculus, where the variables represent the key concepts that characterize physical objects and their state in space-time, for example, mass and shape, coordinates, velocity, and acceleration. The principle of conservation of energy and the law of universal gravitation are added to the theory as axioms. Of course, a scientific theory must be consistent in terms of the mathematical concepts it formalizes, i.e., it must not contain logical contradictions. However, this is not enough. It must also be valid, i.e., no theorem should be falsified by observations.

I explained that our ability to understand phenomena also depends on the arsenal of models available to us. We have often devised new concepts and models, which while contrary to "common sense," simplify the study of phenomena. Examples include the introduction of the concept of infinity to mathematics or the use of the Dirac delta function in physics.

4.1.2 Stratification—Abstraction Hierarchy

The physical world has breadth and depth. Phenomena can be studied at scales ranging from 10^{-35} m, which is the order of the Planck length, the smallest measurable length, up to 10^{26} m, which is the size of the observable universe. In order to understand these from the very small to the very large and in order to overcome the complexity of this 10^{61} change of scale, we study the world at various layers of *abstraction*. Please note that is just a methodological simplification and it does not at all mean that reality is stratified! In contrast, I believe that reality is a single whole—however, this is an ontological question.

First, let me explain what *abstraction* is, as it must not be confused with ambiguity and indeterminacy. Abstraction is a holistic approach to breaking down complexity, highlighting essential characteristics of a phenomenon for each level of observation. Thus, when I observe the world with the naked eye and I am studying physical phenomena or building artifacts, I can forget that the matter is composed of particles; I can even consider that a solid, material body is a point where its entire mass is concentrated. The use of abstraction has proved essential in order to tame complexity.

We use model hierarchies in every field of knowledge, where each layer is associated with the superior layers through an appropriate abstraction relationship. As we climb up the abstraction hierarchy, the scale of observation increases.

Figure 4.1 presents an indicative hierarchical stratification for physics, informatics, and biology. Readers may be more familiar with that of physics, which ranges from the particle world to the universe. It goes through atoms, molecules, and theories of solids, liquids, and gases, which are useful to study complex electromechanical systems that make up artifacts.

Similar stratification can be carried out for computing systems, which in their physical form are integrated circuits implementing the logical gates and memory of computers. The upper strata are networked systems that make up the cyber-world.

An obvious requirement when stratifying knowledge is its unification, in other words, to define relationships between the models at each layer so as to relate the results and achieve a global understanding and knowledge of the field being considered. This would lead to what certain physicists call a "Theory of Everything." Such a unification of knowledge has not yet taken place in any of the key fields noted above. In the case of physics, for example, the general theory of relativity would have to be linked to quantum mechanics, which is known to be an open problem.

The integration of stratified knowledge comes up against insurmountable problems that will be analyzed in what follows.

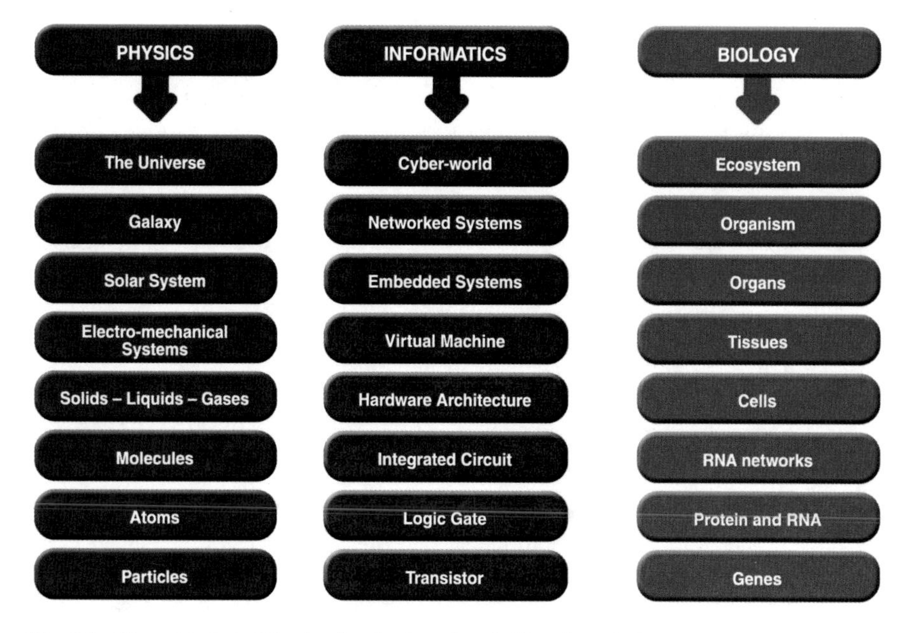

Fig. 4.1 Hierarchical stratification for the three key fields of knowledge

4.1.3 Modularity—The "Atomic Hypothesis"

In each layer of an abstraction hierarchy, to break down the complexity of the observed reality, humans make the very useful hypothesis that the world is built from components, building blocks, composed based on laws that we try to discover. Democritus was the first to have this idea and made a very interesting argument in favor of the "atomic hypothesis" concerning nature. In other words, he tried to prove, through logical reasoning, that the physical world is not continuous but discrete.

Of course, the ontological question "is nature discrete or continuous?", which cannot be approached rationally, is less important. What is important is the fact that the "atomic hypothesis" allows humans to break down complexity and develop scientific knowledge, i.e., to understand phenomena rationally.

The development of physics and chemistry is the triumphant proof of the effectiveness of such a *modularity assumption*. Matter is made up of a small set of elementary particles that, through successive combinations, result in the complex universe. We construct buildings by combining building materials such as bricks and concrete blocks. We manufacture electrical circuits by connecting resistors, capacitors, and inductors. We build machines by assembling parts. In every case, theory allows us to predict the behavior of the whole, by composing the behavior of its components. The composition of components can also be defined in a mathematical or systematic manner. Thus, the behavior of a circuit is defined by the behavior of its components and the way they are connected.

We should also note that natural languages evolved according to this building principle: the meaning of a phrase is a synthesis of the meanings of its words.

As a methodological approach, which considers a complex system to be the composition of a relatively small number of types of components, the *modularity assumption* is based on the following three basic rules:

1. The whole is a composition of a finite number of types of components. For example, an atom consists of electrons, protons, and neutrons; similarly, a chemical compound is a combination of different kinds of atoms. A cell is a synthesis of cell organelles. A phrase is made up of words.
2. We can learn the characteristics of each type of component by studying it individually, just as we can separately examine the behavior of particles, atoms, or the meaning of words.
3. Our knowledge of the whole can be inferred by synthesizing our knowledge of the components that make it up and the place of the components in the structure of the whole.

It must be noted that if, thanks to the second rule, we can learn the behavior of each type of component individually and know how the whole is structured using different types of components, then by applying the third rule we can know the behavior of the whole.

For example, by knowing the behavior of the components of an electrical circuit (voltage source, resistor, capacitor, and inductor) and the connections, we can study the behavior of the circuit. Our knowledge regarding each type of component is equations that associate current with voltage. The connections state how we can use the equations of the components to determine a system of equations characterizing the behavior of the circuit.

Note that in this example, the synthesis of global knowledge assumes the locality of interactions between the components making up a system. Each component interacts with components in its "neighborhood." This assumption of the locality of interactions does not appear to hold true in quantum physics or in biology, which makes the relevant phenomena much more difficult to model and understand.

Furthermore, the application of the third rule requires that the behavior of a component we have learned by studying it separately does not change when we synthesize it with other components or, at the very least, that it changes in a predictable way. Unfortunately, this hypothesis does not hold true for information and biological systems, nor for natural languages. It is well known that the meaning of the words of a phrase depends on context, and this makes it extremely difficult to translate natural languages. Furthermore, informatics lacks satisfactory results that allow us to know the properties of a software by synthesizing the properties of its components.

This discussion shows that modularity that was so helpful in our understanding the physical world does not carry out in fields of knowledge such as informatics and biology. It applies even less to economics and social sciences, where systems are composed of components, for example, individuals, whose behavior is difficult to study in isolation without taking into account their position within the structure of the system. Human behavior depends so much on the context in which it takes place that it is practically impossible to characterize.

In conclusion, I would point out that when a system consists of a large number of components, but the behavior of each component remains unchanged, however much the number of system components changes, then we can study the behavior of the system using statistical methods. This is the case with the kinetic theory of gases, which studies gases as ensembles of particles that obey the laws of Newtonian mechanics.

However, the statistical approach can only be applied to "disorganized complexity" phenomena involving a very large number of equivalent components, as the economist Friedrich August von Hayek noticed [2]. It cannot be applied to phenomena of "organized complexity" observed in biological, information, or economic systems. This is because, as I explained above, the behavior of each component changes dynamically depending on its position in the system structure. The study of behavior and the control of systems of organized complexity are one of the most important challenges in science.

4.1.4 Emergent Properties

We have explained how stratification allows for a piece-by-piece understanding of phenomena. Each layer is characterized by a specific abstraction level with its own basic concepts and laws. Going up the hierarchy, some concepts may not be applied, while others emerge as important. Thus, stratification may introduce gaps in the understanding of the phenomena under study.

I have already said that a very important problem in every field of knowledge is the consolidation of knowledge between different levels of abstraction: can we explain the properties of components at one layer from the properties of the layer immediately below?

For example, can we know

1. the properties of water molecules from the properties of the oxygen and hydrogen atoms and the rules governing their composition?
2. the properties of an information system from the properties of components of its hardware, properties of the programs being executed and the initial memory state?
3. the properties of our mental processes from the properties of the brain's neural systems?

These questions are of a similar nature and a satisfactory answer is most probably unlikely. Without getting lost amidst technical explanations, I would say that there is a problem of scale. When we move from one layer to a higher layer, new properties "emerge" due to the interaction of a large number of components that form much more complex behaviors as a whole. Of course, just saying that new properties "emerge" does not explain anything. We are simply acknowledging our inability to consolidate the theories of a field of knowledge.

For instance, we know that water consists of a set of molecules and that each molecule is a synthesis of oxygen and hydrogen atoms. Even if I could theoretically determine the behavior of a molecule, there are properties that are the result of the interaction of molecules. A 250 ml glass of water contains 8.36×10^{24} molecules. The emergent properties of water are that it evaporates at 100 °C, freezes at 0 °C and that volume changes according to temperature. Even if it were theoretically possible to faithfully describe the dynamic behavior of molecules and their interactions, modeling them would be extremely complex and practically impossible to resolve.

Similar considerations apply to both other examples. When we view a computer as an electrical circuit described by systems of differential equations, it is practically impossible to go up the abstraction hierarchy and achieve a match between the state of the circuits and the execution of the software.

Matching the properties of the brain with the properties of our mental functions seems to be even harder. Unfortunately, there are scientists who are "over-optimistic" about the possibility of overcoming these difficulties. For example, the Human Brain Project, which received over 1 billion euros in funding from the European Union [3], promised that by studying and simulating detailed models of the brain, it is possible to understand phenomena concerning consciousness. Squandering money on infeasible but "catchy" is a common feature of major research programs.

Our inability to explain the emergence of properties is discussed by Philip Anderson in his article entitled "More Is Different" [4], of which I quote a passage below, because it very clearly explains that our ability to reduce the whole into simple laws does not imply that, starting from those simple laws, we can once again find all the properties of the whole.

> *The main fallacy in this kind of thinking is that the reductionist hypothesis does not by any means imply a 'constructionist' one: The ability to reduce everything to simple fundamental laws does not imply the ability to start from those laws and reconstruct the universe. In fact, the more the elementary particle physicists tell us about the nature of the fundamental laws, the less relevance they seem to have to the very real problems of the rest of science, much less to those of society.*

4.2 Issues Related to Knowledge Application

4.2.1 *The Limits of Scientificity*

As I said, many believe that there is a clear and absolute distinction between scientific and non-scientific knowledge. We must not, of course, overlook the differentiation in the validity of knowledge, but an absolute distinction between scientific and non-scientific is completely arbitrary and dangerous.

I explained that the knowledge is valid by degrees. Even in the natural sciences, for example, astronomy, the possibility of experimental validation and repeatability is limited. We must also distinguish between experimentally confirmed relationships (laws) and the interpretations we normally give for the sake of generality and consolidation. Thus, in the past we believed that space was full of "*aether*" or in the existence of "phlogiston," a fire-like element contained within combustibles and released during combustion.

What often strikes me is the confidence with which renowned scientists discuss grand theories. They make a clear effort to impress the general public in order to gain publicity, popularity and, consequently, financial support for their research.

Every time a theory is proposed, the criteria that permit its refutation must also be given.

There are theories considered to be categorical truths that, when examined according to logical and methodological criteria, give rise to questions and doubt. Such theories are those trying to explain what happened in the past based on present-day observations, such as cosmological theories and theories on the evolution of species. The formulation of these theories raises a particular logical problem.

While, in order to discover physical laws, we study a cause-and-effect relationship in the here and now, in the case at hand, we do know the outcome, but we have to guess the root causes. We must perform a logical operation called abduction, which, unlike deduction that proceeds from the hypothesis (cause) to the conclusion (effect), seeks to find the "most likely cause." The difference with deduction is that many hypotheses can logically lead to the same conclusion.

Some people liken this logical problem to the problem of a detective who arrives at a crime scene and has to guess murder scenarios only from on-the-spot observations, without interviewing witnesses.

Cosmological theories are based on certain hypotheses, which we consider valid as long as they have not been refuted. Such hypotheses are that the universe is isotropic and that the physical laws that we observe apply uniformly with the same values of universal constants and, in particular, that there are no faster-than-light communications.

Generalizing theories validated in our "neighborhood" to apply to the entire universe has a certain amount of arbitrariness. Perhaps we should be slightly more careful and modest when trying to solve cosmological problems, particularly when the prevailing cosmological theory has significant weaknesses. How pleased can one be with a theory, which posits that only 5% of the universe contains matter and energy such as that found in our solar system, the remaining 95% being dark matter (27%) and dark energy (68%)? The entire Big Bang narrative leaves logical questions unanswered, and I do not know if and when these questions might be given a satisfactory answer.

However, even Darwin's theory of evolution, which everyone now accepts as an indisputable truth, has certain logical gaps, which must not be suppressed. First, I would like to state that I do not question the idea of evolution, which is clearly confirmed by observations. However, what needs to be explained scientifically as a possibility is how a single-cell organism can evolve into multicellular organism with eyes, ears, and other organs. How, by applying the basic idea that through a succession of random changes and choices of the fittest (natural selection), we can reach multicellular structured organisms such as humans. I believe that Darwin's idea of natural selection explains the adaptability that can be achieved through learning processes. Technically speaking, it is a process of changing parameters in order to optimize a criterion, which in this case, concerns the viability of the organism. However, what remains technically unexplained is the creation of complex organisms through this "game" of random chance and consequent adaptations to the external environment.

Another way of exploring the problem is to think of it using an analogy: comparing the construction of an artificial system to the development of a biological system. Those who design systems or write software will understand precisely what I am talking about. Optimizing a structured system already in existence is one problem and constructing a system by composing components that add new functionality and cooperate harmoniously is another altogether. Everyone understands that it is not possible for a radio to become a television set by applying a mechanism such as natural selection, no matter how we imagine interaction with the environment.

From this analogy, we can reasonably conclude that biological systems differ radically from artificial ones, and that no satisfactory explanation will be found as long as we limit ourselves to the organism/environment "game" without taking the very properties of organic matter into account. One such property overlooked by Darwin's theory is that organic matter, and only organic matter, is not only able to learn, which largely explains adaptability, but can also self-organize. In my view, these properties that we have to discover make the game of evolution less random.

Perhaps studying genes alone is not enough in order to understand building mechanisms and the emergence of new functions. We must try to describe the "grammar" of the construction of these biological systems, i.e., structuring principles of components at the various layers of the bio-hierarchy. As is the case with other fields of knowledge, the search for theories that provide a holistic "bottom-up" explanation risks proving exceptionally fruitless and disappointing.

4.2.2 Scientism

One implicit hypothesis of physics that, to some extent, can be confirmed experimentally is that each factor that affects a phenomenon is observable and measurable. This does not apply to other fields of knowledge. For example, social phenomena cannot be understood by examining relationships between measurable quantities. Thus, the human factor is frequently ignored in economic models, simply because its behavior cannot be rigorously modeled.

The successes of physical sciences and their technical applications, particularly during the nineteenth and twentieth centuries, gave rise to a philosophical movement, which considers that only mathematically formulated knowledge is worthy of trust and recognition. This led to the unchecked use of mathematical models in many fields of knowledge with mediocre to completely disastrous results.

The trend named *scientism* in the mid-twentieth century concerns the uncritical application of models and approaches from the exact sciences. A typical example of scientism is the failure of economists to formulate successful policies because of their propensity to emulate the physical sciences as closely as possible. Thus, the correlation between aggregate demand and total employment is the only one for which we have quantitative data and will therefore be accepted as the only causal connection that exists [2]. A theory based on measurable data can be considered more plausible than another one that provides certain well-founded explanation, which, however, one can challenge due to the absence of measurements.

We often liken the search for knowledge to the behavior of the drunkard who, as the old joke goes [5], has lost his car keys and is searching for them under a lamppost. A policeman sees him and helps him look together. After a short while, the policeman asks the drunkard whether he is sure he lost the keys there, and the drunkard replies that, no, he lost them in the park. The astounded policeman then asks why he is searching for them under the lamppost, and the drunkard answers "because that's where the light is."

This happens more often than you would think in the search for knowledge. Researchers prefer to write publications about beautiful theories that are far removed from reality rather than racking their brains over difficult practical problems. Few take the trouble of judging the validity of their hypotheses and the applicability of their results. This is a phenomenon that I have also observed in informatics, where beautiful theories propose models that over-simplify reality.

It is not surprising that scientists awarded the Nobel Prize in Economics are allegedly mediocre at business. Econometricians reduce problems to equations and ignore the human factor, which is also the main parameter in an economic game. They ignore the qualitative. The exact same economic policy would produce very different results when applied in two different countries, for example, Germany and Greece.

There are serious reasons to be concerned about the dangers of blithely adopting approaches that are only ostensibly scientific. The progress achieved in the physical sciences over the last decades has exceeded every expectation and any discussion about the existence of limits could be considered suspicious.

Public opinion fails to understand that some ideas promoted under the guise of science may be at best empty, if not misleading, and therefore dangerous. There are countless examples in every field of knowledge.

I remember being an informatics student in 1972, when the Club of Rome published a famous text, in the name of an ill-defined "systems science," entitled *The Limits to Growth* [6], which predicted a pessimistic future for humanity, an immediate shortage of energy resources and raw materials, and their impact on industrial development and the environment. The text was published as a book, sold 30 million copies, was translated into more than 30 languages and remains the top best-seller of books on the environment. I remember that, at the time, they lavishly financed research projects for the development of software simulating economic systems. The conclusions of the text were widely publicized by the media and seriously influenced economic policies around the world; conversely, very little was heard of the radical criticism rightly voiced by authoritative scientists and analysts. Of course, when we now examine what happened at the time, it is clear that the supposed analysis capabilities of then-mainframes were used for economic and political purposes.

It is imperative that we protect ourselves from scientism, which derives its authority from methodologically unsubstantiated analogies to the physical sciences. The risk of controlling social processes in the name of supposedly science-based predictability—especially at present with the use of artificial intelligence—is absolutely real.

4.2.3 *Experts—The Mystification of Expertise*

Scientism often goes hand-in-hand with a dangerous trend of mystifying expertise, which holds that all serious matters must be judged and ultimately decided upon by experts. This mystification is compounded by the management of knowledge by computers.

Today, there are experts on every issue, ready to advise and guide us. They range from those who would advise us on personal problems to those who propose ready-for-use solutions to major social and national issues, for example, how to overcome the economic crisis, how to reduce inequality, and how to address the refugee issue.

This is not to say that we do not need experts to shed light on technical aspects of the complex problems facing societies. However, the solution to every major problem also has a political and moral dimension. When the time comes for choices, public opinion, once properly informed, must be able to take part in making crucial decisions.

I believe that it is possible to explain to the public in a comprehensible way what is at stake in a decision involving risk and what its consequences are. What is necessary is an effort for popularization, for appropriate awareness raising and preparation through education and public debate. Access to knowledge and the development of a critical attitude in the many is essential in order to enable them to participate actively and knowingly in processes and decisions that determine their future.

If crucial decisions are made without transparency by an exclusive circle of experts, advisers, and politicians, this poses a danger to democracy and liberties.

It is interesting to compare how the coronavirus crisis has been managed in different countries. At first, the experts did not agree with each other on the magnitude of the risk and the precautions that had to be taken. However, what made a difference was not so much the knowledge of experts as much as decisions of a political and moral nature. Certain countries made it a priority to save lives over other criteria such as not disrupting the economic life of the country. By this, I mean that experts define a framework of options and the state makes decisions based on value criteria (see Sect. 8.2.2.2—Risk Management Principles and Their Implementation).

Knowledge has long been the preserve of the few. Since antiquity it was used to prop up the powerful, who could buy knowledge and use it as they saw fit. I must point out that education is set up in such a way so as to favor those who memorize, affording little acknowledgment to those possessed of creative and inventive minds. It does not foster curiosity and creativity. It favors intellectual indolence and conformism. At the same time, I am not certain that the possibilities afforded by the Internet to access knowledge have been exploited to the full by providing appropriate services and incentives.

Experts today play an important role in shaping public opinion and directly influence political choices. Acting either as independent professionals or through famed consulting firms and think tanks, they constitute an institutional ecosystem that affects economic and technological development. They usually arrive with in-depth analyses in hand—replete with countless references, graphs, and reports—either to endorse certain political choices or to throw up a smoke screen and muddy the waters.

In many countries, experts lend the authority of expertise to disastrous policies and serve as a necessary complement to the status quo. As a Greek, I experienced first-hand the terrible economic crisis which broke out in Greece in 2007 and which, under the advice of famous academic and institutional experts, led to the application of a series of reforms and austerity measures that proved to be disastrous for the country's economy. It is now more than clear that the country underwent a harsh treatment that resulted in dramatic impoverishment and loss of income and property that compromised the country's independence and mortgaged its future.

In addition to its validating role, the experts' ecosystem may also be parasitic, depending on the environment. I have read studies by experts proposing changes for Greece that were hollow, if not malicious. I have seen committees of experts invested in writing reports, spending countless hours and funds to produce findings that, in the best of cases, end up in the waste bin. Of course, even when they reach specific conclusions and proposals, these prove convenient for those who commissioned them.

The mystification of expertise and the alienation of public opinion from knowledge, which is evolving rapidly and is becoming increasingly difficult to understand, is a danger for modern societies.

4.2.4 Researchers, Research, and Innovation

I will end this chapter on the development of knowledge by saying a few things about organized research and researchers.

4.2.4.1 Research and Researchers

Scientific research is currently aimed at the systematic production of knowledge. It is essential for the development of a modern, innovative economy. It is conducted at universities and research institutions, as well as laboratories of large companies.

The profession of researcher differs from other professions, as it requires strong motivation and dedication, as well as a certain talent for creativity. What I found wonderful in this profession is the sense of freedom, a prerequisite for exploring new pathways for creation. However, managing this freedom while possessed of a sense of responsibility is not a simple problem. I have seen researchers lose themselves in choices leading to impasses, as well as researchers who prefer safety in taking beaten paths, which are free of risk but where there is little chance of breakthrough results. A professor used to tell me that "you will not invent the light bulb by trying to improve on candle-making."

Throughout my life, I tried to carve out my own path by following a vision of my own. At times, I achieved this, and at others, I did not. I have tried to keep myself outside mainstream thinking, with which I did not agree, and that has had and still entails a certain cost.

There are currently large communities of researchers, each working on specific issues, who share a common concern and who try to push the boundaries of knowledge as far as possible. Research is less free than it was 60 years ago. It is driven by large programs that are generously funded by states and businesses and aim at what we call scientific and technical challenges. The programs are multi-annual and each absorbs significant research funds. That is why it is often discussed whether and to what extent the funds spent on a program will deliver the expected results within a reasonable period of time. Without elaborating on this very important topic, I would stress that there must be strong social control over the priorities and directions of research.

I raise the issue in order to underline the difference between a well-defined scientific and technical challenge, on the one hand, and a vision, on the other. A challenge has as its objective to overcome an obstacle in the development of knowledge, and this objective has the following characteristics:

1. The objective is relatively well defined and concerns: either achieving a specific goal such as proving a difficult mathematical result and developing a vaccine or seeking a framework that consolidates existing results and proposes a new approach to the development of knowledge, for example, the unification of theories in physics.
2. The objective is realistic in the sense that it takes into account the state of the art in the field of knowledge in question, as well as theoretical and practical limitations, for example, complexity and experimental evidence, respectively.
3. Achieving this objective will constitute a breakthrough in the development of knowledge. In other words, it will bring about a significant qualitative change in knowledge that is different from everyday gradual progress.

With this characterization, I wish to make a clear distinction between challenges and visions. A vision also represents a long-term ambitious goal, but it does not meet any of the above criteria. For example, one such vision was the "Strategic Defense Initiative," the so-called Star Wars programme, launched in 1983 by US President Ronald Reagan [7]. It was clear from the outset that the objectives of this program were not achievable and I think no expert ever believed otherwise. Although the program failed to achieve its original objectives, it mobilized investments in the defense sector that may have yielded significant "collateral" results.

Today, the AI vision mobilizes the massive investment and the enthusiastic involvement of big tech companies. Nonetheless, there is a lot of confusion about what intelligence really is. The media and big tech companies, with large-scale grandstanding projects, spread fallacious opinions suggesting that human-level AI is only a few years away. The excitement and buzz about the advent of a new intelligence era are amplified by the predictions of various "experts" about a future inevitably dominated by machines. This does not allow a sober, informed, and reasoned debate to ponder the risks of the AI revolution that most people tend to accept as a fatality we cannot escape.

The problem with visions is that, beneath visible high and fancy objectives, they may hide other objectives, unspoken and less noble.

What are the chances that the work of CERN, the European Organization for Nuclear Research, which has an annual operating budget of one billion euros, could achieve the goals it has set? These include pushing "the frontiers of science and technology for the benefit of all" and helping to "discover what the universe is made of and how it works" [8]. It is difficult to imagine how adding to the already long list of discovered particles, new particles can contribute to achieve these grandiose goals.

What needs do space programs serve? Why would we want to colonize other planets? And even if this is possible on a small scale, the costs would be unbearable and the benefit small. A manned journey to Mars involves enormous technical difficulties—and yet the media often talk of colonizing this planet as if it were possible in the near future.

"Visionaries" often invoke "progress in scientific knowledge" regarding all the above. However, the issue at hand is one of the priorities and hierarchy of objectives. There are more urgent challenges raised by environmental and epidemiological problems at the global level that do not cause the same kind of enthusiasm. It would therefore be good if the compelling visions asserted by large organizations and maintained by powerful lobbies with the support of the media were analyzed in depth according to the criteria mentioned above.

Research is costly today. It makes no sense to squander huge sum of money simply to satisfy some curiosity or to achieve "progress in knowledge" in a general, vague sense. Research must contribute to the well-intended progress of humanity and have clear objectives that do not run counter to the common good.

4.2.4.2 Research and Innovation

The emergence of innovation as a driving force of economies is a relatively recent phenomenon. The research and technology model that developed after World War II changed radically in the 1990s, mainly for two reasons.

First, there was an acceleration of the innovation cycle, where basic and applied research worked in tandem, particularly under the pressure of competition and the market. Second, the development of basic research became a luxury, feasible only for very large companies.

Today, businesses are less vertically integrated. In order to minimize costs and risk, and in cooperation with research institutions, they develop research and technology of common interest that they tailor to meet their own needs. In such partnerships, venture capital plays an important role in exploiting research results by creating start-ups.

Collaboration between businesses and universities has radically changed the structure of research at universities and the criteria for evaluating researchers. We have transitioned from small theoretical research teams to large laboratories, where hundreds of researchers collaborate on ambitious research programs.

I remember, as a child, picturing a researcher, wearing a white coat and thick glasses, hunched over a laboratory desk, working night and day to come up with a magical formula. Today, we are far removed from this model of knowledge production. We need significant funds and infrastructure, as well as *critical mass* and *excellence*.

Critical mass is essential because research is closely linked to technology and, therefore, there needs to be cooperation between basic and applied research. Furthermore, research is often interdisciplinary and must be linked to relevant fields of knowledge.

We also need excellence, which means the capacity to generate knowledge and apply it to innovative ideas. The creativity of a talented researcher cannot be replaced either by financial resources or by masses of inexperienced researchers.

The production of innovation is based on the creation of *innovation ecosystems*, which are the result of synergy between three factors: (1) research institutions, (2) industrial enterprises, and (3) start-ups. The most illustrious examples are to be found around major technology universities in the USA such as Stanford, Berkeley, MIT, and CMU.

An innovation ecosystem brings together significant human resources from pools of experts, scientists, engineers, and managers and is characterized by a special culture of creativity. It is attractive due to the cultivation of a special lifestyle—see the case of Silicon Valley. The state intervenes by providing incentives and planning, ensuring access to foreign technologies, special tax regimes, and copyright protection. Financial institutions—venture capital funds and banks—support start-ups. Of course, everyone works together in order to produce innovative products and services.

The admirable complementarity of the roles in the synergy between the three actors in an innovation ecosystem must be emphasized. Research institutions carry out basic research and work with industrial enterprises to implement it. They thus secure research funds from their industrial partners and enjoy contact with the new problems brought about by technological development. Start-ups are often founded by researchers and inventors. They have the advantage of flexibility and efficiency compared to large industrial enterprises.

Innovation is not the privilege of superpowers and large countries such as the USA, Japan, and Germany. Small countries, such as Israel, Switzerland, and the Nordic countries, are among the top in the innovation arena. Each of these countries has developed innovation ecosystems and robust research structures based on the advantages and strengths gained over the last decades. The most striking cases are certainly Israel, a technological superpower, as well as Switzerland, which has a very "aggressive" policy for the purchase of patents and rights.

In conclusion, I would like to emphasize the importance of the human factor for building an innovation-driven economy. Lack of vision cannot be offset by money. Policies should support creative individual initiative, favor meritocracy, and fight corruption. They should implement measures fostering entrepreneurship and attracting innovative businesses. They should reform research institutions to help them achieve critical mass and excellence and provide incentives for connection to the real economy.

There are countries with huge scientific potential that are doomed to fail simply because their policies privilege economic and geographic determinism and ignore how much the human factor matters in the innovation race.

References

1. https://en.wikiquote.org/wiki/Ludwig_Wittgenstein
2. https://www.nobelprize.org/prizes/economic-sciences/1974/hayek/lecture/
3. https://en.wikipedia.org/w https://en.wikipedia.org/wiki/Human_Brain_Project iki/Human_Brain_Project
4. https://www.tkm.kit.edu/downloads/TKM1_2011_more_is_different_PWA.pdf
5. https://en.wikipedia.org/wiki/Streetlight_effect
6. https://en.wikipedia.org/wiki/The_Limits_to_Growth
7. https://en.wikipedia.org/wiki/Strategic_Defense_Initiative
8. https://home.cern/about/who-we-are/our-mission (accessed on 08/01/22)

Part II
Computing, Knowledge and Intelligence

We analyze the relationship between computational processes and physical phenomena, as well as the production and application of knowledge by machines and humans.

- We investigate relationships between informatics and physics and discuss how each of these fields of knowledge can enrich the other, comparing basic concepts and models.
- We attempt to compare artificial and natural intelligence, trying to answer the question of whether and to what extent computers, as they stand today, can approach human mental functions.
- We present autonomous systems that are called upon to replace humans in complex operations by realizing the vision of strong AI and discuss the risks—real or hypothetical—of the reckless use of these systems.

Chapter 5
Physical Phenomena and Computational Processes

5.1 Scientific Knowledge and Computing

The models of the physical world and those of informatics share a common feature that can serve as a basis for comparison between them: they are dynamic systems with states and actions that constitute transitions from one state to another. However, there are also very important differences in the way we develop knowledge and the way we formulate and interpret it. One difference already explained is that scientific knowledge is empirical, while our knowledge of computing systems is a priori, at least in terms of theory of computation.

In what follows, I will explain the similarities and differences between computational processes and physical phenomena. Of particular interest is the concept of natural computers leveraging physical processes since it adopts a completely different computation model that is not subject to the intrinsic limitations of the conventional algorithmic model.

5.1.1 Computational Processes and the Experimental Method

The models of computation, as defined by A. M. Turing, are discrete, as they are based on arithmetic. Computation is the process where a machine executes an algorithm by performing a finite number of actions (steps) in order to calculate the value of a function for given values of its arguments. When the process terminates, the state of the machine provides the result.

Machines are modeled in terms of a mathematical relationship that determines how the machine transitions from one state to another. A state consists of the values of the data that the machine transforms when executing the algorithm. I have already explained that the result of the computation is independent of what is happening in space-time. One could define a concept of "logical time" as the number of steps a

J. Sifakis, *Understanding and Changing the World*,
https://doi.org/10.1007/978-981-19-1932-9_5

machine has performed after it started executing an algorithm. However, this process is sequential, and the concept of logical time is independent from physical time.

I have pointed out that our knowledge of computing machines is based on mathematics and therefore different from our knowledge of physical phenomena. In order to understand the behavior of a machine, the experimental method applied to develop scientific knowledge is precarious.

Let us assume that we have a program and want to check if it can compute the function $f: x \rightarrow x^2$. In other words, when I input the number x into the program, it will output the square of x as the result of its execution. Let us perform an experiment where for values of $x = 0, 1, 2, 3\ldots$, I receive the result of the computation $f(x) = 0, 1, 4, 9\ldots$ For what number of correct experimental data can I reasonably reach the generalization that the program actually computes the square of x? Why does the inductive method of generalizing observations, which applies to physical phenomena (see Sect. 3.4.3.2—On the Nature of Scientific Knowledge), makes no sense for algorithms and, therefore, for the programs that describe them?

This is because simple physical phenomena are robust, in the sense that small changes have isometric effects. This does not apply to algorithms due to their discrete nature. Changing slightly in memory can change their behavior radically.

However, since algorithms are defined as mathematical models, we can—at least in theory—achieve stronger guarantees through mathematical verification, that is to say, to prove a theorem based on logical analysis of the algorithm's code, that for each value of x it will compute x^2.

Thus, contrary to physical phenomena, understanding computational processes involved in the execution of programs requires logical analysis of their behavior. This concerns proving theorems by using the mathematical definition of the programming language of the programs and axioms about its data structures. Such an analysis may lead to the discovery of invariant relationships (invariants) linking program variables that remain true over program execution. Therefore, each program is a separate world governed by its own laws, which are indirectly defined by its programmer. However, they are different from physical laws, as they are immutable and can be uncovered through logical analysis and proof.

5.1.2 Physical Phenomena as Computational Processes

There are two different approaches to correlating physical phenomena and computational processes.

The first approach, followed by *digital physics*, considers that the universe can be modeled as a computer endlessly executing a program [1]. The states of this program are defined by the values of the physical quantities observed. The executed program runs through states that precisely characterize the changes to physical phenomena. Without going into technical details, there are various ways of digital representation, one of which—and which I find interesting—uses cellular automata: see Stephen Wolfram's *A New Kind of Science* [2]. A cellular automaton is a potentially infinite set of simple machines ("cells") arranged in space so that each machine can locally interact with neighboring machines. While the functioning of each cell can be very simple, a cellular automaton presents strikingly complex emergent behaviors—see *The Game of Life* by John Horton Conway [3].

The second approach, rather than considering the universe to be a computer, interprets physical phenomena as computational processes that change the values of physical variables according to physical laws.

This leads to the concept of a *natural computer* in the following way. Let us assume that I know the law governing a physical phenomenon—for example, I know that if I throw a stone from a certain height, it will follow a parabolic trajectory. We can consider that in this case, the stone is a computer that solves the equation of parabola in real time.

This very simple idea that every physical phenomenon also involves a computational process proves to be very fertile. It was initially applied to the so-called analog computers, which were used to solve linear systems of differential equations.

Analog computers are simple electrical circuits whose behavior is described by systems of linear differential equations. We can therefore use them to study any problem that involves solving such equations. All that the "programmer" needs to do is to build an electrical circuit where each variable of the problem is represented by voltage at a point in the circuit. Thus, the solution to the problem is the observed behavior of the circuit: for each variable of the problem; it is given by the function describing the change in voltage at the corresponding point.

At present, we have other interesting examples of natural computation. These include neural networks, which mimic the processes of the neural circuits of the brain, and are used effectively for machine learning; these were previously discussed under Sect. 3.3.2. There are also quantum computers and bio-computers that leverage properties of quantum phenomena and of proteins, respectively, to implement basic sets of operations.

Natural computers do not execute programs written by programmers. They effectively compute families of functions whose parameters we can define. They "mimic" physical systems and are best suited to model physical processes, including learning, as proven by the achievements of artificial intelligence.

5.2 Comparing Physical Phenomena and Computational Processes

5.2.1 Computation and the Concept of Continuum

Physical reality cannot be understood without the concept of space-time. Time is the common parameter of the observed quantities, which change in a "synchronous" way as its values increase. Each point in space-time defines a "now" and evolution from each point takes place in parallel as time progresses.

Regardless of the nature of the phenomena of the physical world—continuous or discrete—the assumption that space-time is continuous has proven extremely useful in classical physics. Without differential equations and integration, it is impossible to understand most phenomena. Continuous models are best suited for mathematical analysis and can describe phenomena that discrete models cannot capture.

Perhaps we would have never discovered the idea of continuity if we saw the phenomena on a different scale, for example, one of millimeters. There are no straight lines in nature. When a drop of water slides down a glass, it will follow a notionally straight trajectory that will be distorted by the irregularities of the surface—after all, there is no such thing as a perfectly smooth surface. The idea of continuity raises logical problems that have been discussed since antiquity. The length of a beach is different if measured by a man, a crab, or a microbe! It can be as large as we want it to be, if it is a continuous quantity. Of course, according to quantum mechanics, the length is discontinuous—as the minimum measurable length is defined by the Planck length. However, this does not detract from the usefulness of continuous models.

An ordered set is continuous if, between any two elements a and b such that $a < b$, there is another, different element c such that $a < c < b$. Such sets are rational and real numbers. Only continuous mathematics allow for faithful modeling of physical phenomena. All classical physics and technical culture in general are based on continuous mathematical models. If we confine ourselves to discrete mathematics, then we cannot study dynamic phenomena that are usually described by differential equations.

The idea of continuity in nature was criticized by Democritus and Zeno with his famous paradoxes. The fable of the race between the swift-footed Achilles and the slow-moving turtle is well known. Every step Achilles takes is half the distance that separates him from the finish line, while the turtle is moving at a steady speed. What Zeno observed is that the turtle will always reach the finish line first—assuming, of course, that space is continuous. The proof is based on the observation that Achilles will always be at a certain distance, however infinitesimal, from the finish line due to the continuity of space.

Continuity is a very useful abstraction. Even if nature contains no straight lines or perfect circles, we use these concepts to understand space and its properties in a formal way. An abstract geometric shape is defined using very little information—a line is defined by two points, and a circle is defined by its center and radius. Therefore, we believe that we are driving on smooth roads with no discontinuities, and that the risk of any potholes is marked with appropriate signs. What makes reality predictable is the smoothness of continuity.

The fact that computers deal only with discrete domains limits their ability to simulate converging sequences of events described only by continuous mathematics. Let me explain this by giving an example.

I want to study the movement of a sphere that I allow to fall to a floor from a certain height. As is indeed the case, every time the ball hits the floor, it loses part of its energy and, as a result, it eventually stops. Describing this phenomenon using high-school-level physics, in order to calculate when the sphere will stop, we need to find the limit of the converging sequence of ground collision times. The difference between successive collision times can become smaller than any finite quantity. No computer can compute that limit.

If we ask the computer to solve the equations that govern the phenomenon, it will stop its computation at a point that is, perhaps, very close to the limit. The obvious reason for this is that it cannot handle infinitesimal quantities that are involved in the computation of the limit and that are smaller than any measurable quantity but greater than zero. The computer has a finite memory. An infinitesimal quantity whose digital representation exceeds its memory capacity can always be found.

A different approach to computing such a limit would be to program the computer to prove the convergence of the sequence. However, in order to do so, we need to apply mathematical induction by using rules of logic and algebraic properties. This is also infeasible because computers cannot systematically discover inductive assumptions because of Gödel's theorems (see Sect. 4.1.1.2, Formalized Languages—Theories).

In summary, I would say that computers, which are based on discrete models that do not "have the power" of continuous models, can only approach the accuracy of the continuous—and this at a significant computational cost. The obvious question here, of course, is how programmers of simulators of physical phenomena can realistically represent their dynamics, for example, in video games or even in real-time control and virtual reality systems.

The answer is that there are techniques devised by people who understand the dynamics of the systems and program them appropriately so as to solve them. However, these techniques are specific to each dynamic system. They cannot be generalized and automated.

5.2.2 *Conflicts and Resources in Systems*

Discreteness of computing begets phenomena, such as the conflict between actions to obtain resources. Such phenomena are of particular interest because they also emerge in economic and social systems, as well as in mental processes, as we will explain in Chap. 7.

We consider that a system is described by: (1) a set of variables, which represent its *states*; and (2) a set of *actions*:

1. The possible valuations of the variables define the set of the states of the system. In the case of a mechanical system, the variables are usually continuous. For example, the state of a moving body is determined by its position, speed, and acceleration. In the case of a program, the variables are discrete and represent the program data. A state is defined by a snapshot of the values they take during a step of execution.
2. An action is a rule governing the change of states. It is possible from a given state only when the values of the variables characterizing the state satisfy certain

preconditions. For example, a thermostat acts by ordering heating to start when the room temperature (the variable) is, for example, lower than 18 degrees. The precondition for an elevator door to open is for the elevator to have stopped at a floor.

The execution of an action, when the relevant preconditions are met, results in a transition from the present state to a new one, which implies a change in the values of the system variables. Therefore, in the case of the action of purchasing a product worth 100 euros using my credit card, the precondition is that the crediting of an amount of that value is allowed. When the transaction takes place, my bank account will enter a new state (reduced by 100 euros).

The concept of a *resource* is common and very important for all kinds of systems. The existence of sufficient resources is a precondition for actions to take place and states to change. For an action, a resource is a state variable for which some minimum value is necessary for its execution, i.e., resources are involved in action preconditions.

A resource may be *quantitative* and may represent physical or economic quantities, such as energy, power, memory, time, physical goods, or money. Therefore, time and memory are the resources required to execute a program, money to buy a product and force/energy to accelerate a body.

Furthermore, a resource can be *qualitative*, i.e., its availability is a precondition that may or may not be fulfilled. For example, knowledge is a qualitative resource that enables actions to be carried out: I either know or do not know how to swim, how to converse in English, how to cook stew, etc. With regard to qualitative resources, we assume that the corresponding variable has a value of 1 or 0, depending on whether it is available or not.

Each action needs a minimum amount of resources to be executed, which, in the case of qualitative resources, is conventionally considered to be 1. When the action is executed, it consumes the necessary resources and may release other resources. For example, when a program is being executed, it will reserve the amount of memory and the processor time necessary for its execution. When the program ends, it releases the memory used. It should be noted that there are resources that are consumed when used, such as program execution time.

In other cases, the resource quantity remains unchanged: resources are committed when used and released after use. These include immaterial resources, such as knowledge, as well as various body organs whose use is necessary for an action, such as movement or problem-solving. For example, when I walk, I use my legs, which are non-consumable resources for the action of locomotion. The chemical reaction $H_2 + O \rightarrow H_2O$ requires two hydrogen atoms and one oxygen atom to produce one molecule of water in the presence of a catalyst, which is a non-consumable resource.

The existence and management of resources is important for the functioning of systems, as they determine the feasibility of actions and their overall behavior. However, there is a key difference between physical continuous systems and computational discrete systems.

In physical systems with continuous actions, if two actions taking place at the same time need common resources in order to be realized, the allocation of resources per action is carried out in a continuous manner, so as to satisfy the basic laws of physics.

Conversely, in discrete systems, such as computers and human transaction systems, it is possible to have a *conflict* between two actions, which can be described as follows:

Let us suppose two actions a and b are possible from the state of a discrete system, and each action needs quantities of a consumable resource, for example, memory or money, which are ra and rb, respectively, in order to be carried out. If the available quantity of resources is less than the sum of $ra + rb$, the execution of an action will reduce the available resource to such a level that the execution of the other action is no longer feasible.

In other words, a conflict between two actions a and b can occur when the two actions need a common resource in order to be executed and the execution of one disables the execution of the other.

There are countless examples of conflicts not only in computers, but also in everyday life. When I have a certain amount of money available which is not enough to fully carry out two actions, then I have to resolve the conflict by making a choice. Similarly, if the execution of two programs a and b by a computer requires computation times ta and tb, respectively, and the computer is available for a time that is less than the sum of $ta + tb$, then executing one program does not allow the other to be completed.

In summary, a system has a set of states and actions. States define the available resources (which are consumable or non-consumable). The execution of each action from a state depends on preconditions for the resources required. When an action is carried out, the system enters a new state where the required resources are consumed, and new resources may be released.

Competition between actions when resources are limited follows a "dog-eat-dog" logic—the action chosen to be executed takes up the shared resources and precludes other actions. When two or more actions conflict, information is needed to choose those that make the best use of resources and avoid deadlocks.

Deadlocks are a very characteristic phenomenon of real-life discrete systems. A system is in *deadlock* if it has exhausted all available resources and no action is longer possible. How can this happen? Deadlocks are a kind of "bankruptcy" of systems.

Let us assume a computer has 100 memory units available for program execution, and that this quantity of memory is sufficient to execute any program. In practice, the computer "runs" several programs concurrently, for example, for graphics management, connection to the Internet, various applications, etc. When a program is being executed, it "asks" the memory management system for an initial quantity of memory, followed by successive memory requests in proportion to its needs, until the program ends. The program then releases all the memory it was allocated to be executed.

It is now easy to imagine how a deadlock can arise: when the computer has a total of 100 memory units and a certain quantity has already been allocated—let us say a total of 90 units—to the programs already running, and each one needs more than 10 units in order to be able to terminate. The only solution in this case is to stop at least one program in execution to free up the memory required for the others to continue being executed. It is clear that deadlocks are dangerous for computers running critical applications because they may lead to the abnormal termination of programs.

However, deadlocks arise not only in computational processes, but also in any system that manages energy, monetary, and even immaterial resources, such as rights. The insolvency of a company is precisely a form of deadlock. Another example of a deadlock in the context of a political crisis is a situation where the current rules do not lead out of this situation—all possible actions are prohibited.

Deadlocks are undesirable situations that block system evolution. This is why resources must be managed with great care. Computers use resource management systems that make decisions based on predetermined rules. If there is a conflict between two actions *a* and *b*, the rules generally aim at allocating resources according to a certain criterion. A "fairness" criterion would require alternation in the resolution of conflicts between *a* and *b*. Other criteria are based on the use of priorities according to the importance of the actions. Thus, the actions of an alarm system have a higher priority than other actions taking place under normal circumstances. This is also the case with a road junction, which is the resource shared by vehicles, where traffic on a central road can take priority over a secondary road intersecting it.

Of course, we can build conflict-free systems by sufficiently increasing the resources available. We then have a state of abundance where there is no conflict between the actions sharing these resources. Such systems do not make efficient use of their resources and are therefore uneconomical: they require the availability of as many resources as the maximum demand from all their actions.

There is another simple category of systems where, by design, there can be no conflict: systems whose state unequivocally determines their evolution. These systems are deterministic, such as the electro-mechanical systems of classical physics. Given the initial state, the subsequent states are generally defined by the time elapsed since the system began. Programs with a single "thread of execution" are also deterministic, i.e., at each step, only a single command can be executed.

In summary, I would say that resource management and deadlock avoidance are very important problems for discrete systems and computers in particular. This is a fundamental difference between continuous physical phenomena and discrete processes that describe computational as well as economic and social phenomena. In the following chapters, we will demonstrate that conflict is a key concept for understanding consciousness and mental processes.

5.2.3 *Time and Synchronization*

Physical phenomena are described by relationships involving variables that are functions of time. Much has been written and said about time. Let us confine ourselves to saying that time is but a common parameter of physical quantities whose states change as the parameter increases, i.e., time passes. An important concept here is *synchronization*: for every single infinitesimal increase in the parameter of time, all physical quantities change continuously and simultaneously. It is just like a movie. The position, speed, and acceleration of a moving body change as a function of time. When we have many moving bodies in the same frame of reference, by knowing their initial kinetic state, their positions are determined only by the time that has elapsed since they began moving. This wondrous property of physical phenomena at the scale we observe them makes understanding them extremely simple. As time progresses, everything takes place in a predetermined and parallel manner within space. This property of determinism, which is linked to the absolute synchronization of phenomena and which we now consider so "natural," is not so easy to capture as a computational process.

The German mathematician and philosopher Gottfried Wilhelm von Leibniz, a contemporary of Newton and co-inventor of differential calculus, wondered how two pendulums with exactly the same characteristics could move in an absolutely synchronous way, without any interaction between their movements. This phenomenon is of course, explained in physics by the fact that the pendulums are suspended in the same gravitational field characterized by g, the gravitational acceleration constant. Whatever the explanation of the phenomenon according to physics, Leibniz was right as a profound thinker and philosopher: what nature achieves without apparent cost has enormous computational complexity because there is no built-in notion of space and time in algorithms.

The theory of computation is based solely on arithmetic and is independent of the physical properties of space-time, which can only be modeled in an approximate way. If we had programs in the place of the pendulums, their exact synchronization would entail considerable computational overhead. Moreover, since computation is discrete, the synchronization would not be perfect as it is in nature.

There are significant differences between physical time and concepts of time we can associate with computational processes. Of course, in both cases, time helps us comprehend change in the phenomena.

We have said that physical time can be understood as a monotonously increasing parameter that "marks" all events in the physical universe. Each event may be dated depending on the observer. Physical time is sometimes called *exogenous time* because it is not possible to stop its passage since change is inherent to physical reality. It could be thought of as a wheel, which as it turns, determines all changes in the phenomena. It allows us to correlate the rate of change of phenomena occurring in parallel in the universe.

We have also said that a program computes a function as a logical sequence of commands and that it is independent of physical time. The execution time of a program depends on the speed of the computer executing it.

However, when we use computers to simulate physical phenomena, physical time is represented by a variable that increases as the computer executes the simulation process. This variable changes according to the simulation speed and can be characterized as *endogenous time* because its speed of change bears no relation to the actual time of the computer. Its value represents the time of the phenomenon being simulated. Thus, simulating a second in the evolution of a complex phenomenon can last days, depending on the speed of the computers executing the simulation program.

Today we use powerful computers to simulate—for the purpose of experimentally studying—the behavior of complex electro-mechanical systems, such as cars or aircraft when they are being designed. In other words, we use virtual models instead of building physical models, even on a small scale. These design techniques afford engineers the advantage of efficiency by both reducing development costs and allowing experimentation, simply by changing the model parameter combinations. This is what we call *model-based design*.

A computer simulation of a complex system with many variables that are functions of time involves the execution of a circular process that computes the evolution of each variable by steps that correspond to the change for a short period of time, a time step Δ. By knowing the values of the system variables (the state of the system), the computer calculates the new values after time Δ has elapsed. However, there are phenomena that change very fast and the time step Δ must be variable to take into account bursts of events. Adjusting the values of Δ to the dynamics of the phenomenon being simulated is a challenging problem.

The simulation becomes even more difficult when the model is so complex that it cannot be resolved by a single computer. In such cases, we have to use arrays of cooperating computers, which requires the synchronization of the simulation programs being run by all the computers. In other words, the endogenous computation times of the simulation programs must coincide as much as possible, and this can entail a computational overhead so large that it can neutralize the advantage of parallel computing.

Note that the physicist Richard Feynman, who is considered the father of quantum computing, discussed this problem in 1982. In his article titled "Simulation Physics with Computers" [4], Feynman says: "*The rule of simulation that I would like to have is that the number of computer elements required to simulate a large physical system is only to be proportional to the space-time volume of the physical system. I do not want to have an explosion.*"

Without going into technical details, I would say that the simulation of physical phenomena by computers involves high complexity due to the discrete nature of their computation and the cost of synchronizing their simulation processes. This is something that physical time achieves in a completely "magical" way, with no cost whatsoever.

5.2.4 *Why Is the Physical World Comprehensible?*

A general observation we can make when comparing the physical world to the world of computers at the level where we observe them is that the physical world has admirable uniformity and normality in its structure and behavior. The same basic, simple laws govern physical phenomena everywhere in space-time. In contrast, each program can be considered to be a universe with its own particular laws. These laws are generally complex and are very difficult to guess experimentally. As we have said, there are techniques that allow us to discover them: given a program, one can theoretically compute invariant relationships between its variables.

If the physical world were built like the world of programs, then there would be no science whatsoever and scientific truths would go out the window.

We have explained that physical laws resulted from a process of generalization, which is logically arbitrary. Humans discovered them experimentally. For example, we found that the ratio between the current and the voltage of a resistor is constant and that the air resistance of a moving body is proportional to the square of its speed. However, as I have explained, this process of observation/generalization in order to discover the laws governing physical phenomena does not work in computing systems. On the one hand, I have a "uniform" behavior captured by simple laws, while on the other hand, I do not have any such behavior.

Why can we not have discrete computer systems that behave "normally," such that, after a relatively large number of experimental observations, we can risk making a generalization with high likelihood as we would do for simple physical systems?

Let us make another comparison. The physical world is a combination of particles, while the computer world is a combination of computational elements. How can it be that, at a high layer of the hierarchy in the physical world, simple and predictable laws emerge, while the same is not true in the computer world? Regarding this wondrous property of the physical world, Einstein said that "*the most incomprehensible thing about the universe is that it is comprehensible.*" The basic laws of physics are astoundingly simple and normal cause-and-effect relationships can be characterized by simple linear laws. This allows us to approximate physical phenomena fairly faithfully using systems of linear differential equations. Of course, there are also more complex and chaotic phenomena. Small changes to certain parameters can change the entire dynamics of the system—the "butterfly effect" comes to mind. However, all technical civilization is based on the fact that the electro-mechanical artifacts we build adhere to extremely simple laws that allow for predictability.

Computing systems, even in their most elementary form, are difficult to understand. The vast majority of such systems are nonlinear, such as the simplest system, which is a memory component. Furthermore, computing systems are, by their very nature, chaotic due to the discrete nature of information. Can we have programs that behave robustly? Can we find a programming language where small changes in the program's code would result in a small difference in their behavior, as is the case with small changes in the parameters of an electro-mechanical system, which generally do not bring about dramatic changes in its behavior?

Nature does in fact possess a robustness that is alien to computers. If, for example, a civil engineer designs a building whose structural stability has been calculated for steel of a certain strength, it is certain that the building will be structurally sound if higher strength steel is used. This does not apply to computer systems, which are not resilient in this way. What we call "anomalies" appear: the improvement of certain parameters of a system's components does not bring about the reasonably expected improvement in the overall behavior of the system.

A well-known example concerns *time anomalies*, where the following "paradox" occurs: When one executes a program on a computer, one reasonably expects that the faster the computer, the shorter the execution time. There are cases where this is not true. Performance not only does not improve but may actually deteriorate.

These anomalies are due to the non-determinism of computational processes and the way in which conflicts are resolved, as discussed above. When we try to improve system performance by increasing the resources available, the degree of freedom in resource management increases and new choices emerge, which could lead to devastating results.

Similar anomalies occur when we increase the available memory. It is known that lack of available memory during the execution of a program can lead to deadlock. However, in certain cases, increasing the quantity of memory in a system functioning normally can also lead to deadlock.

This discussion indicates that there are significant differences between physical and computing systems. These differences are due to the discrete nature of the latter, whose behavior and properties present interesting similarities to economic and social phenomena.

References

1. https://en.wikipedia.org/wiki/Digital_physics
2. https://writings.stephenwolfram.com/2017/05/a-new-kind-of-science-a-15-year-view/
3. https://en.wikipedia.org/wiki/Conway%27s_Game_of_Life
4. http://research.physics.illinois.edu/DeMarco/images/feynman.pdf

Chapter 6
Human vs. Artificial Intelligence

The relationship between human intelligence and artificial intelligence is a very topical and much debated issue due to recent advances in machine learning and its spectacular applications in the development of smart services and systems.

6.1 Human Intelligence Characteristics

6.1.1 Slow and Fast Thinking

It is a well-known fact that our mind combines two ways of thinking [1]: (1) *slow and conscious thinking*, which is procedural and applies the rules of reasoning, and (2) *fast and automatic (non-conscious) thinking.*

We use slow thinking to consciously solve a problem, for example, to analyze a situation, to plan an action or to build an artifact. Fast thinking is automatic and allows us to solve problems whose complexity is extremely high, such as when we talk, walk, play the piano, etc. When we walk, our "fast computer" solves a problem whose algorithmic real-time solution would be extremely complex. Understanding and analyzing fast thinking is extremely challenging. If a pianist playing a difficult composition uses conscious thinking in an effort to analyze exactly what she is doing at a given time, she may become confused and lose her pace.

Automatic thinking is the most important factor in human intelligence. If we compare the information processing capacities of the two systems, we see that a large part of its bandwidth is due to fast thinking. Of course, this comparison is quantitative and does not reflect the role and importance of conscious thinking for humans.

There is also remarkable cooperation and complementarity between the two modes of thinking. Following birth, the conscious and slow system invents processes that are automated by the fast system. When an infant babbles and is consciously saying "ma-ma" the fast "computer" learns progressively until it takes control of the function of language. The same happens when we learn how to ride a bicycle. First,

J. Sifakis, *Understanding and Changing the World*,
https://doi.org/10.1007/978-981-19-1932-9_6

we consciously try, following a "trial and error" process, to balance on the two wheels. Ultimately, the fast "computer" learns how to balance automatically, without us even understanding the laws of mechanics.

There is a stunning correspondence between the two modes of thinking and the two basic models of computation. Conscious thinking is based on mental models that we understand and is what we use to program computers. Automatic thinking is the result of learning, just like the computation of neural networks.

If we want to build a bipedal robot or a robot riding a bicycle, we can take two different approaches: we will either program the computers controlling the movement of the robot, or we will use a suitably trained neural network. In the first case, we must use the theories of mechanics and write the real-time control programs. In the second case, we must train the neural network to learn how to balance, exactly as we learn without using theoretical knowledge.

Mathematics and logic, being creations of conscious thinking, reflect how it works. The algorithms and programs that computers execute are but formalized processes of this mode of thinking.

Incidentally, I would like to stress that what we call "causality" is not necessarily a property of physical phenomena. It can simply reflect the way we understand phenomena because conscious thinking is serial, and the rules of reasoning formalize preservation of truth as a cause-and-effect relationship. Furthermore, computational processes involve causal chains of operations where, when an operation is completed, its results trigger subsequent operations.

Of course, in mathematics we can formulate relationships, that are not causal, and therefore not executable as a single computational process, but that characterize sets of computational processes that are capable of multiple dynamic correlations.

Natural computers that rely on physical processes directly, and neural networks in particular, are best suited to model fast thinking. They are inherently parallel and can perform computation whose analysis using logical processes is practically impossible. Unfortunately, fast thinking is non-conscious and, therefore, understanding and analyzing its basic laws and mechanisms would be a hopeless endeavor.

An interesting question arises: Which functions can natural computers effectively compute? Is it possible to have a new theory of computation that uses the computational capabilities of physical systems, for example, analog, neural, quantum?

6.1.2 Common Sense Intelligence

I have always said that any "intelligence" that conventional computers possess reflects the intelligence of their programmers. They simply execute commands describing the mechanical processing of symbols.

Of course, things have changed with the advent of neural networks, which follow a radically different computational model that is not programmed but simply learns from very large datasets. However, we certainly have "smart systems" that specialize in solving one type of problem: they can play chess, classify images, take part in television game shows, etc.—but a system that is a chess grandmaster will fail badly at driving a car.

One step toward approximating human intelligence would be to have computers that exhibit what we call strong AI. In other words, they could combine solutions to specific problems and skills, exactly as people do, by responding to stimuli from their environment.

What characterizes human intelligence is the combination of perception/interpretation of sensory information, their logical processing and decision-making that could potentially lead to action. Such behavior has nothing to do with the development of strategies for playing chess or go. In games, the rules are predetermined and do not change. Humans are able to adapt to environments where rules and goals change dynamically depending on circumstances.

As I will explain in what follows, we can understand consciousness as our mind's ability to "see" ourselves acting in a *semantic model* of the external and internal world. This model is, at first, built automatically from our infancy. It is then consciously enriched through learning. It can be seen as a dynamic system with states that reflect, on the one hand, our perception of the external environment and, on the other hand, our awareness of our internal state. Without such a semantic model, it would not be possible for us to understand languages and therefore communicate with each other. At the same time, the use of language plays an important role in correlating the semantic models each of us develops individually.

Much of human intelligence is due to what we call *common sense*. Our minds can use their semantic model to assess a situation and the consequences of what is taking place in our environment. The model is created by the accumulated experience and is almost automatically enriched every day. Thus, the "parent–child" relationship is enriched by a set of other relationships that common sense uses, such as age, hierarchy, and support, which are very difficult to list and formalize.

For computers to exhibit this type of behavior, we must endow them with a corresponding semantic model. In theory, we could build such a model if we could analyze and formalize natural languages; if we could model them with hierarchically structured semantic networks of relationships between concepts, equipped with rules for representing and updating knowledge. For example, when defining the "parent–child" relationship, we have to imagine and formulate all the relevant relationships and rules it entails. Unfortunately, despite research efforts for more than 50 years, very little progress has been made in this direction.

For example, humans interpret the sequence of images shown in Fig. 6.1 almost instantaneously as an aircraft accident. This is because the process of understanding can link the contexts of the perceived sequence of images using common sense knowledge of the semantic model. In contrast, a computer can analyze the sensory information for each image but lacks the knowledge that would allow it to reach the same conclusion. It can identify and perhaps correlate the objects appearing in a single image, but it is unable to analyze and understand the dynamic relationship between images.

Fig. 6.1 Humans are endowed with common sense reasoning

6.1.3 Cognitive Complexity—The Boundaries of Understanding

Understanding means to be able to link an observation to a relationship one has mentally mastered. We can realize the proportionality between force and acceleration or the exponential growth of an organism over time because we have already mastered the relevant concepts and their properties at school or through experience.

When we study phenomena affected by a large number of parameters, we are forced to make certain simplifications based on assumptions that some parameters are primary and others are secondary. Such simplifications are not always legitimate. For example, to consider phenomena without friction in many cases simplifies theory. There are cases, however, where friction is essential to the dynamics of a phenomenon and cannot be ignored.

There are phenomena that are inherently complex and cannot be simplified, such as meteorological, economic, and social phenomena. The number of parameters that need to be understood is quite large. Therefore, we cannot study these phenomena in theory, not because they cannot be causally understood, but because the discovery of relations between observations involves an inherent complexity that the human mind cannot grasp.

It is well known that the human mind is constrained by *cognitive complexity*, which we define as the time needed to grasp a relationship. It has been experimentally determined that the limit to the number of parameters that our mind can correlate is approximately five (one relationship with four parameters). This hindrance to conceiving relationships with many operands is a significant constraint that affects our ability to understand complex phenomena, but also limits the complexity of the theories and artifacts that we can build. I noted above that Einstein was right to observe that, for our good fortune, the basic laws of physics are astoundingly simple. The existing scientific theories involve a relatively small number of independent quantities and concepts to formulate these laws. I can also say from personal experience that complex theories are difficult to understand and therefore difficult to use, while checking their validity often proves problematic.

Therefore, developing theories that would allow a holistic approach to the study of complex phenomena comes up against the constraints of the human mind. It is here that I believe cooperation between humans and computers could help us overcome these restrictions, to a certain extent, as I will explain in what follows, after discussing the validity of machine-generated knowledge.

6.2 Weak Artificial Intelligence

Living organisms and humans, in particular, can be considered to be computing machines that bear certain similarities to computers: they use memory and languages. The correspondence between hardware/software and brain/mind is also noteworthy.

However, there are also some important differences. Computational phenomena in living organisms are "resilient"—they have inherent adaptability mechanisms and, most importantly, have allowed for the emergence of concepts and language.

Computers exceed human thinking in being able to compute faster and more accurately. This enables them to compete with humans in solving problems that require the examination of an unfeasibly large number of solutions or a combination of a large number of predetermined data. This idea is applied in systems such as IBM's Deep Blue and Watson and Google's AlphaGo. The fact that human intelligence is defeated in such games leads some people to believe that computers are "smarter" than humans.

It is worth briefly discussing the evolution of artificial intelligence, which was born in the mid-1950s as a branch of informatics that "studies and designs intelligent systems." Since the mid-1960s, its proponents would enthusiastically promise that they could build machines that would rival human intelligence, as two examples of declarations from that period show: "machines will be capable, within 20 years, of doing any work a man can do"; "within a generation ... the problem of creating 'artificial intelligence' will substantially be solved" [2]. Of course, such promises were not kept for many reasons. One is due to a lack of understanding of the limitations of the theory of computation. Another, more profound reason is that in order to build a machine that behaves like a human, we must have understood and analyzed the mechanisms of human intelligence.

The focus and perimeter of artificial intelligence has shifted over the years, with corresponding fluctuations in the interest and funding of research. There are two characteristic periods known as "AI winters": one in the early 1970s and one in the late 1980s.

Since the beginning of the twenty-first century, there has been a rekindling of interest, particularly thanks to achievements in machine learning and data analytics. This is accompanied by a "frenzy of optimism" regarding the importance and impact of artificial intelligence, particularly on the development of autonomous systems and services.

Many still consider that intelligence is limited to solving complex but well-defined problems by focusing on decision-making. They believe that, thanks to the use of machine learning methods, they will be able to rise to the challenge.

I believe that what characterizes human intelligence is autonomous behavior and adaptation to changes in the internal and external environment. It is the incomparable ability of the human mind to create new knowledge, to understand situations one has never encountered before, and to set new goals. The gap between human and artificial intelligence will be bridged when we have computerized systems that autonomously perform a large number of services while adapting to changing situations.

6.2.1 *The Turing Test*

In order to judge whether a computer *A* is just as intelligent as a person *B*, Alan Turing proposed a test. The test consists of having their behavior compared by an interrogator *C*, who sends questions to *A* and *B* and who, in turn, provide a corresponding answer for each question (Fig. 6.2). If the interrogator cannot tell from the answers, which is the computer and which the person, then it is concluded that they are equally intelligent.

The relevance of the Turing test was challenged by philosopher John Rogers Searle, who proposed the Chinese Room Argument. This is a thought experiment, which shows that programming a computer may make it appear to understand language by manipulating strings of symbols without understanding their meaning.

The experiment, the layout of which is shown in Fig. 6.2 to the right, assumes that *B*—whether man or computer, it does not matter—is in a room receiving a number of questions written in Chinese by *A*. The experiment intends to prove that there is a

Fig. 6.2 The Turing test (left) and the Chinese room argument (right)

way for B to give C the right answer to every question asked by A without speaking a word of Chinese. Therefore, the Turing test, as well as all other tests comparing behaviors, is not appropriate.

Let us assume that B has access to a huge database containing all possible questions that can be worded in Chinese and their respective answers in the same language. This is theoretically feasible, as there is but a finite number of questions in Chinese.

Thus, when B receives a question, B will search the database, find the corresponding answer and give it to C. It is obvious that this process can be automated by ordering the questions lexicographically. This thought experiment shows that, if we take observed behavior as a benchmark, we may be led to erroneous conclusions.

Another argument concerning the Turing test is that a set of specific questions may prove to be the undoing of a human participant. For example, if the interrogator asks "What is the hundredth digit of the decimal expansion of π?", the computer can respond immediately, while the human will be unable to answer relying solely on his or her mental faculties. Thus, the computer is superior to humans when facing this type of question!

Note that we understand the world by observing and studying behaviors. Therefore, the only *methodologically legitimate* criterion is comparing behaviors, and we have seen that this is not fit for purpose.

However, if we use ontological criteria, a distinction is feasible. Without knowing exactly what the mind is, we can say with certainty that the computer is not a mind because the mind can build a computer, whereas the opposite is impossible given the present circumstances.

A closing methodological remark that will be further discussed in Sect. 6.3 concerns autonomous systems. The criterion for intelligence cannot be reduced to a Q&A game; it should be about building systems capable of replacing human agents performing tasks in complex organizations. This is a much more ambitious goal than the one set by the Turing test.

6.2.2 Machine Learning and Scientific Knowledge

Recent advances in machine learning and data analytics have shown that it is possible to generate knowledge and forecasts with some success. Machine learning and statistical analysis techniques allow for the recognition of complex relationships in data. These correlations may indicate cause-and-effect relationships and allow for predictability. However, they may also be random, which is difficult to establish, but I will not discuss the matter further [3].

How valid is the knowledge generated by neural networks and how much can we trust it? A comparison of the way it is generated, and the way scientific knowledge is developed allows us to highlight key differences.

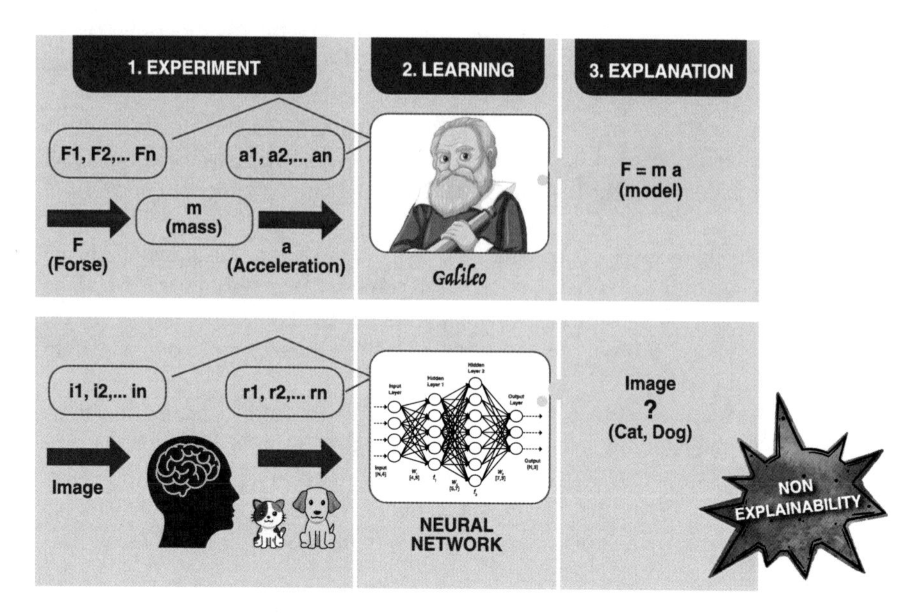

Fig. 6.3 Processes for producing scientific and machine learning knowledge

As we have said, the scientific method enables the development of knowledge that explains our observations regarding the world. Scientific discovery is the result of the learning process of an observer, who formulates a hypothesis by creating a model that allows for the explainability and predictability of a phenomenon.

On the other hand, neural networks can generate knowledge as the result of a long "learning" process involving a large dataset. By learning the data representing a cause-and-effect relationship, the network can estimate the most likely result for a given cause by applying a kind of extrapolation method. The success rate depends on the degree of "training" of the network, but we cannot be certain that the response is correct. Furthermore, our inability to use models in order to understand exactly how the network works does not allow us to estimate the possibility of an incorrect response.

Figure 6.3 shows differences between the scientific approach to studying a physical phenomenon (acceleration a of a mass m from a force F) and the process whereby a neural network learns a cognitive process (how to recognize images of cats or dogs). In both cases, there is a common goal: to capture the input-output function characterizing a cause-and-effect relationship. In the first case, it is acceleration a resulting from a given force F, while in the other, it is "cat" or "dog," depending on what the input image depicts. In both cases, there is a learning stage at first.

In the physics experiment, the experimenter, let us say Galileo, learns by relating cause and effect. By using his capability of mental abstraction and creativity, he will assume proportionality in this relationship. This hypothesis will ultimately become a "law" since it is experimentally validated.

Using a mathematical model allows checking the validity of knowledge (see Sect. 3.4.2). We can use the model to examine the physical phenomenon and see how it behaves in *corner cases* and to generally gain insights, thanks to mathematical analysis.

Similarly, machine learning involves an initial experimental process, where images are labeled by an experimenter. Then the neural network is "trained" by adjusting its parameters so that for each image i_n it provides the correct response r_n. The difference with the scientific approach is that we cannot characterize the input-output behavior of the neural network through a mathematical model. This would require, for this particular example, an almost impossible formalization of the concepts of cat and dog.

However, note that when the network inputs and outputs are physical quantities, there is no theoretical limitation in this respect. By knowing the structure and behavior of the components of the neural network, I can theoretically calculate the mathematical function that characterizes the input-output relationship.

Neural networks are particularly useful where mathematical models cannot help. This is where mathematics finds itself at a loss, as we do not know how to define theoretically, what an image means. This is just as difficult as formalizing natural language.

Nonetheless, the ability of neural networks to deal with non-formalizable knowledge gives rise to an interesting complementarity with conventional computers that will prove very fruitful in the future.

6.2.3 A New Kind of Knowledge—Prediction Without Understanding

Can computers help us overcome the limitations that our human nature imposes on the search for knowledge? The answer is clearly yes. There are numerous examples where computers contribute to the analysis of complex phenomena, thanks to the combined use of data analytics and machine learning. It is perhaps, thanks to the use of computers, that we will be able to overcome the barrier of cognitive complexity, which was discussed under Sect. 6.1.3. Computers may help us develop and verify new theories to explain complex phenomena.

Thus, we have a process for generating a new type of scientific knowledge, where the law is not a clearly formulated mathematical relationship devised by the human mind, but a possibly complex relationship discovered with the use of computers. The analysis of this relationship may allow predictability, but it will certainly limit the deeper understanding of phenomena that would allow their characterization by explicit mathematical models.

It is, therefore, thanks to computers that we have a new type of knowledge that allows prediction without the understanding enabled by the power of mathematical analysis.

The use of artificial intelligence and supercomputers is paving new roads for the development of knowledge. This was the argument made years ago during a discussion with a seismologist acquaintance of mine, who told me that perhaps in a short while Google would be better at predicting earthquakes than experts. I do not know whether this will come to pass. However, I think that new techniques for analyzing large datasets can increase the possibilities of predicting complex phenomena, without necessarily understanding their nature in the absence of explicitly formulated laws. Taking earthquakes as an example, if we could correlate seismic activity in one location to seismic activity around the planet, then perhaps we could make predictions without or with very little theory.

Of course, we must weigh the impact of using such knowledge. The "cloud" is becoming a type of oracle that people turn to in search of solutions to the complex problems they face—is this good? bad? dangerous?

The generation and use of knowledge through artificial intelligence techniques that allow for predictability without understanding, especially if used to make critical decisions, should give us pause.

6.3 Beyond Weak Artificial Intelligence—Autonomy

6.3.1 Autonomous Systems

I have explained that what distinguishes computers from humans is that current computers automatically execute certain predetermined functions, whereas humans can act autonomously; that is, they can act reactively to changes in their environment and also act proactively, driven by internal goals.

Today, the use of artificial intelligence allows us to take an important step forward, to progress from automated systems to autonomous systems that are called upon to replace people in complex functions. The main motivation is to achieve greater efficiency by combining already automated processes without human intervention. The human factor will only regulate certain goals, delegating their realization to autonomous systems. For example, in the case of self-driving cars, we will simply enter a destination; in the case of smart factories and farms, we will simply enter production indices.

Achieving autonomous systems and services is a core goal of the Internet of Things. If this goal is achieved, then we will have much more convincing proof of a certain kind of computer intelligence that is far more important than beating people at games.

At this point, the difference between automation and autonomous systems must be explained. To this end, I will present five different systems in order of difficulty of design: a thermostat, an automated shuttle, chess-playing robots, football-playing robots, and self-driving cars (Fig. 6.4).

The common characteristic of these systems is that they use computers to control their environment, so that their behavior meets certain goals. Computers receive

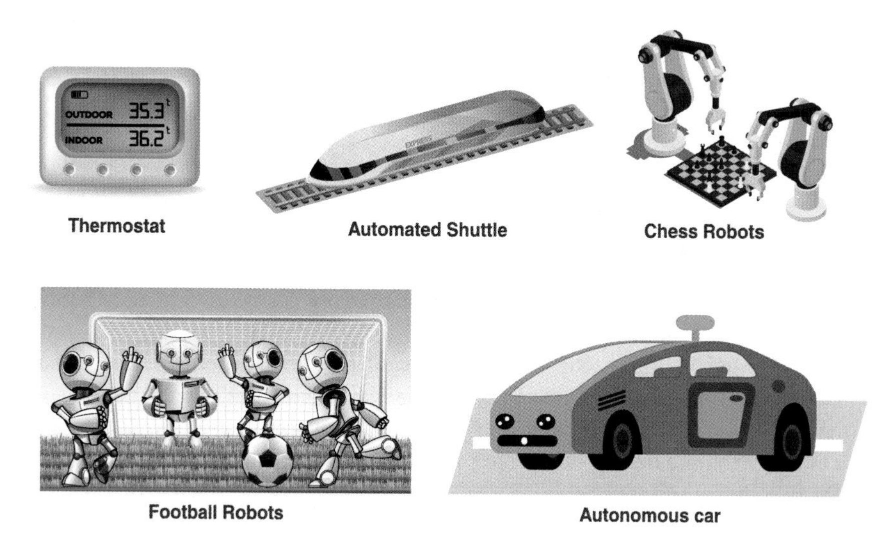

Fig. 6.4 Automated and autonomous systems

information about the state of the environment via sensors, and compute commands, which they send to actuators that perform the appropriate actions to achieve the goals.

A thermostat controls the operation of a heater in order to keep the temperature of a room between a maximum and a minimum value. When the temperature given by the sensors reaches the minimum value, it commands the heater to begin functioning. When the temperature reaches the maximum, it orders the heater to stop.

An automated shuttle has a more complex control system. A system that enforces a series of predetermined stops at stations with corresponding acceleration and deceleration rates. The system takes into account sensor signals that determine the position of the shuttle along its path. Designing such a system does not pose any particular difficulty in its core principles. Particular attention should be given to how to accelerate and decelerate in order to ensure the safety and comfort of passengers.

A chess-playing robot has a relatively simple environment, whose state is determined by the position of the pawns on the chessboard. However, its control system is extremely complex and impossible to design, similar to the two previous systems. The reason is the astronomically large number of combinations on the chessboard, and the even larger number of moves for each combination in order to achieve the game's objective. Consequently, movements cannot be determined statically (a priori), but are calculated dynamically. For each state, the robot uses pre-existing knowledge to calculate tactics, i.e., sequences of actions, which, depending on the reactions of the opponent, lead to the best result.

A football-playing robot faces an even more complex environment defined by the players' positions and speeds. A major difference from the previous example is that the environment changes dynamically. Therefore, the football-playing robot must be able to monitor changes in the environment so as to react successfully in real time. This means that images from its cameras must be analyzed precisely and in time in order to depict the state of the pitch as faithfully as possible. Furthermore, the robot control system dynamically calculates its goals according to its role and position on the pitch. For example, there are goals that involve defensive behavior and others that involve offensive behavior. For each specific goal, the robot calculates the respective tactics over a certain length of time, taking possible reactions by its opponents into account.

Finally, self-driving cars are certainly the most complex system for many reasons. First, the physical environment is changing dynamically and is not limited to a pitch or a chessboard. Moreover, the number of all the vehicles and the position of the obstacles around it is also changing. The layout of the environment depends on geography and the infrastructure available (traffic regulation sensors and equipment, communication networks). Finally, each vehicle has an extremely complex control system. It uses computers that manage multiple goals to shape a strategy consisting of selecting the most important goals first for safety and second for comfort. When it selects a set of compatible goals, it calculates corresponding tactics to achieve them.

The above comparison shows the significant differences between automated and autonomous systems that are in a way, called upon to demonstrate mental faculties commensurate to those of humans. The thermostat and train are simply automated systems because they perform predetermined services with fixed goals in a well-defined environment.

The other three systems are characterized as autonomous, because they have the capacity, similar to living organisms, to generate and manage knowledge with a dual purpose: on the one hand, to "understand" their external environment and, on the other hand, to adaptively manage multiple goals and plan corresponding actions to achieve them.

6.3.2 Characteristic Functions and Organization of Autonomous Systems

Following the above comparison, I propose below an architecture for an autonomous system that clearly shows how five key functions work together to achieve autonomy. A similar model can also be used to reach a theoretical understanding of how the basic mental functions of consciousness can be combined (see Sect. 7.2.2).

An autonomous system accepts and processes sensory information from the environment and computes commands whereby actuators carry out actions that change the state of the environment. For example, in Fig. 6.5, the system is the autopilot of a self-driving car. The internal environment is the vehicle whose direction and speed are controlled by the system. In the external environment, we see three vehicles, one pedestrian and one traffic light. The system processes the information concerning the environment, both internal and external, and issues

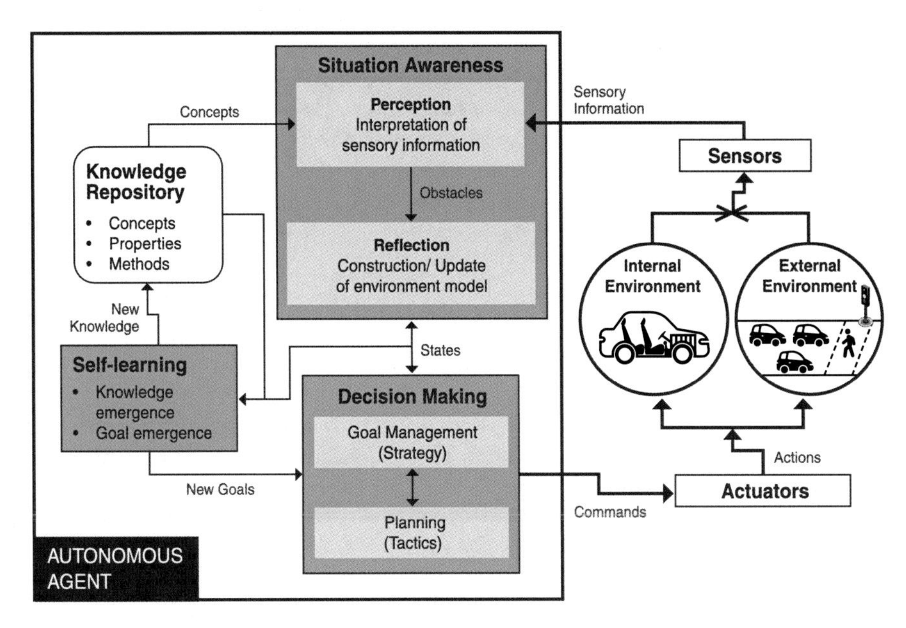

Fig. 6.5 Architecture of an autonomous system with its five key functions

commands to carry out actions. These actions must take place within predetermined time limits defined by the dynamics of the environment in order to achieve the goals in time.

An autonomous system combines five key functions, two of which aim at understanding environment situations (perception and reflection) and two at decision-making (goal management and planning). The fifth function confers the ability of self-learning.

The system is also equipped with a *knowledge repository*, where it stores acquired knowledge that is useful for identifying and managing sensory information, in particular. Knowledge includes, first and foremost, concepts concerning objects in the environment and their properties, as well as methods for decision-making. In the example at hand, the concepts of "car," "pedestrian," and "lights" are required to "understand" the external environment. The repository for each of these concepts may contain information concerning their characteristic properties so that the system achieves better predictability, for example, it may know the maximum speed and acceleration of a specific type of car.

The function of *perception* receives sensory information from the environment (image, signals) and analyzes it by distinguishing between concepts and, possibly, relationships linking them that are stored in the repository. Therefore, in our example, the sensory information from the external environment contains three cars, one pedestrian, and one traffic light with corresponding information on their position and their kinetic state. Regarding the internal environment, the sensory information concerns the kinetic state of the vehicle, such as speed and acceleration.

The function of perception is usually accomplished using neural networks, which are currently the only technology fit for this purpose.

The perceived information is transferred to the *reflection* function, which is tasked with building a model of the external and internal environment of the system. This model has state variables that represent states of the environment such as the kinetic attributes of the obstacles and the state of the vehicle. The actions of the model are changes in the states that the controlled vehicle or the obstacles around it can perform.

The model must be updated in real time so as to reflect the dynamic state of the environment as closely as possible.

The environment model is used by a decision-making module that combines two functions.

The first function performs *goal management* by selecting, from a set of pre-defined goals, a subset of compatible goals in relation to the current state of the environment model. Goal management determines what we call the system's *strategy*.

The system's goals are divided into *negative* and *positive goals*.

Negative goals concern the avoidance of undesirable states, such as safety goals regarding collision avoidance.

Positive goals concern the achievement of desirable conditions, such as optimizing passenger comfort, fuel consumption, and moving from one location to another.

We can also distinguish among short-term goals, such as safety goals, medium-term goals, for example, for maneuvering the vehicle to overtake other vehicles or drive through intersections, and long-term goals, such as completing an itinerary.

Timely and critical goal selection is crucial to system autonomy because it is highly complex and requires a computation time that must satisfy real-time response requirements.

Goal management is complemented by the *planning* function, which determines the system's *tactics*. For each set of selected goals, this function calculates a sequence of commands to the actuators, which carry out corresponding actions to realize these commands. Thus, with regard to the collision avoidance goal, it must control speed and direction by adequately combining braking, acceleration, and steering wheel angle. For every kind of maneuver, the system must have the appropriate tactics to achieve its goals.

Finally, the fifth key function is *self-learning*, which manages and updates the knowledge on the repository. Knowledge is updated through the creation of new knowledge concerning: (1) the environment, for example, new concepts based on accumulated knowledge from the analysis of model data; and (2) new goals for adapting to changes in the environment or changing parameter values that are relevant to the choice between goals.

The self-learning function is a key feature of human autonomy. When someone has never driven on snowy roads, they can adapt their behavior by carefully testing various tactics and with the goal of minimizing risk. The self-learning potential of actual systems is limited to parameter optimization but cannot create completely new concepts or new goals.

Note that the functioning of the system is cyclical. The cycle begins with perception and continues with reflection that updates the environment model. This is followed by decision-making, possibly with the choice of new goals, and the implementation of tactics not been completed during the previous cycle. This is because the cycle duration must be short enough to achieve some short-term goals (of the order of a tenth of a second for safe driving), while achieving long-term goals may take millions of cycles (to reach a destination). Obviously, when new goals are selected in a cycle, they must be compatible with those already selected and not yet realized.

The above architecture defines autonomy as the capability of a system to achieve a set of coordinated goals without human intervention, adapting to changes in the environment. In order to achieve autonomy, it is necessary to combine the five mutually independent functions: perception, reflection, goal management, planning, and self-learning.

This characterization allows us to distinguish between automated and autonomous systems.

A thermostat is an automated system because it does not need any of the five key functions.

A shuttle is primarily an automated system although modern versions use the function of perception to analyze images.

The other systems can be considered as autonomous, but with graduated difficulty.

For the chess-playing robot, the perception and reflection functions are relatively simple, as the environment and its possible changes are slow and well defined. Decisions are also made based on clearly defined rules. The difficulty lies in computing successful tactics anticipating the opponent's tactics.

For the football-playing robot, the perception and reflection functions are quite complex due to the dynamic nature of the sensory information. The strategy also dynamically changes in cooperation with the other robots on the same team. The goals may be offensive at times and defensive at others, depending on the robot's position in the game. The development of tactics through cooperation between the players is also dynamic. In order to coordinate the functions, the use of knowledge is important and includes the rules of the game, as well as knowledge gained dynamically by learning the characteristics of the players, especially those on the opposing team.

Finally, autonomous cars are among the most difficult systems to achieve due to the complexity of their environment and the need to manage multiple goals and adapt in real time.

6.3.3 Should We Trust Autonomous Systems?

The development of reliable autonomous systems is essential in order to realize the vision of the Internet of Things. It is both a scientific and a technical challenge, with tremendous economic and political stakes. That is why all major technology companies are investing in this sector, focusing particularly on self-driving cars, because of the considerable economic incentives. This battle is being fought by companies such as Google and its subsidiary Waymo, Apple, Intel and its subsidiary Mobileye, Uber, the Chinese companies Alibaba, Huawei, Tencent and, of course, all the major motor manufacturers, with Tesla at the forefront.

The solutions promoted by industry mainly rely on the use of neural networks, as the algorithmic techniques performed by conventional computers seem to be unable to solve the problem of perception effectively.

Here I must point out that the use of neural networks in critical autonomous systems has been the subject of fierce debate, as it raises serious safety problems. Until recently, the construction of critical systems had to be based on scientific knowledge and the use of mathematical models. When we build a bridge or design the autopilot of an aircraft, the use of mathematical models allows us to predict how these systems behave in various scenarios and to guarantee their safe use with a very high probability, for example, for a civilian aircraft, the non-hazardous failure rate per hour of flight will be less than 10^{-9}. I must also stress that all artifacts are developed on the basis of regulations and standards controlled by independent certification bodies. When you buy a toaster or car tires, an authority has tested and guaranteed, based on theoretical and experimental data, that if you use them correctly, your life will not be at risk.

Unfortunately, machine learning systems are not based on models, but on "accumulated" empirical knowledge. We can, of course, ascertain experimentally whether they are working properly. However, even if they work perfectly according to a large body of experimental data, we cannot say with the confidence afforded by the scientific method that they will continue to work properly (see Sect. 6.2.2). The certification methods applied by independent organizations require explainability that cannot be achieved by machine learning systems.

Today, in order to avoid stopping the development of autonomous systems with neural networks and to maintain their national supremacy, the competent US authorities accept their use after "self-certification." In practice, this means that the manufacturer—and not an independent authority—guarantees their safety and is held fully liable in the event of an accident. Thus, if the autopilot system of a car fails, the manufacturer must pay compensation for any damage and victims.

This policy change, which does not impose objective safety criteria and allows every manufacturer to act freely, poses serious risks. Human life could simply become the adaptation parameter of an equation, where economic and technical criteria are set off. In my view, the use of autonomous systems should be limited when we cannot guarantee an adequate reliability level.

6.4 Artificial Intelligence—Threats and Challenges

I will not be discussing the possibilities offered by the use of computers and artificial intelligence, in particular. They are numerous and relatively well known. The media often talk about them and the radical changes in the way we live, work, and learn. The automation of processes and services offers the advantage of efficiency. Without direct human intervention, we can have "real-time" control, so to speak, of the use of resources, in sectors such as energy, telecommunications, and transport, in the best way possible so as to achieve economies of scale and quality of life.

I will discuss the risks, whether hypothetical or real, in greater depth.

6.4.1 Hypothetical Risks

Mythology has recently grown around the issue of computer "ultra-intelligence." According to one version, computer intelligence will eventually exceed human intelligence and perhaps we will end up as something like pets for machines.

These views on the risk posed by the intelligence of machines are espoused by celebrities such as Stephen Hawking, Bill Gates, and the entrepreneur and inventor Elon Musk. Some are influenced by the ideas of Ray Kurzweil that the technological singularity is imminent, which is supposed to happen when the computing power of the machines outpaces the "computing power" of the human brain [4].

Obviously, all these arguments are flimsy and lack seriousness. Having very powerful machines is not enough to overcome human intelligence. These ideas find a platform in the media that disseminate them uncritically and resonate somewhat with the public. I believe that the scientific community must react to this mixture of obscurantism and buzz and propose a sober assessment of the prospects opened up by artificial intelligence based on scientific and technical criteria.

Unfortunately, people enjoy believing in thrilling stories and imaginary dangers. Public opinion is easily impressed by hypothetical threats, such as alien invasion or the end of the world according to the Mayan calendar. In contrast, understanding real risks and responding to them as a result of a calm and rational analysis always comes too late (e.g., the foreseeable economic crisis which, nevertheless, crushed us). Science fiction literature is full of stories of evil robots and aliens that threaten humanity.

I quote the three basic rules of ethics which, according to the renowned science fiction writer Isaac Asimov, robots must observe: (1) do no harm humans, (2) obey humans, and (3) protect themselves.

Of course, the computing systems we use are not the "evil robots" of science fiction. However, no one questions whether these systems can breach any or all of these rules with enormous consequences—and, of course, we cannot blame them of bad intentions, as is the case in scientific fiction stories!

6.4.2 Real Risks and Challenges

Perhaps the buzz about the purported intelligence of computers conceals other, real risks. This is where problems arise, particularly of a social and political nature, concerning the type of social organization and the relationships they serve.

6.4.2.1 Unemployment

A danger identified long ago regarding ever-increasing automation is unemployment for professions where the use of robots will eventually come to prevail. Thus, we will see a gradual drop in the number of jobs in sectors such as agriculture and industry, as well as services that can be automated. There will be employment in occupations that require intensive creativity, for example, system programming and design, or that require no qualifications and are not easy to systematize, for example, mail distribution. This trend, in addition to the high unemployment rate, will broaden the gap between well-paid jobs requiring skills and knowledge, and other manual occupations.

Of course, some people argue that the disappearance of jobs will be balanced out by the creation of new needs. I believe that the problem of unemployment and the wage gap will deteriorate unless radical reforms are effected on the structure of occupations—a subject I will not discuss in further detail.

6.4.2.2 Security, Safety, and Risk Management

The lack of security of information systems can prove extremely dangerous when the degree of automation integration exceeds a certain threshold. It is well known that it is not possible to achieve total protection against cyber-attacks, even for the most vital systems. At best, we can hope to detect intruders in time. Unfortunately, for technical and other reasons, computer systems will remain equally vulnerable in the foreseeable future. This means that we cannot rule out disasters, particularly those taking place during times of tension between countries. Today, cyber warfare is not waged by individual hackers, but by well-organized businesses and even states.

A related safety risk is the interdependence of solutions offered by an increasingly complex technical infrastructure built in an empirical manner—like the pyramids. We know that we cannot change complex software written in old languages, such as Fortran. Unfortunately, large computer systems are not always structured in a modular manner like electro-mechanical systems. It is not easy to disconnect some of their components and replace them with equivalent or even better ones without disrupting their functioning. Such changes involve risks that are difficult to assess.

This difficulty of replacement and evolution makes reliance on certain initial options binding for the future. For example, if you were to switch from right-handed to left-handed traffic, the cost and risk would be enormous. Currently, Internet protocols do not possess the desired characteristics in terms of security and reactivity. However, initial choices commit us with regard to their fundamental characteristics.

I have already discussed a significant difference in risk management that was introduced by artificial intelligence and autonomous systems. There are no longer independent state institutions that guarantee and control the quality of systems and their safety. This responsibility is passed on to the manufacturers! The risk is obvious, since the user safety level will be determined not by technical and transparent criteria, but by an optimum between manufacturing costs and insurance costs to cover accidents.

Unfortunately, the current massive adoption of information and telecommunications technologies is seen as a one-way street. No one is asking what kind of technologies we should develop and why, or how to use existing technologies in the most appropriate way. There is no public debate on the economic, social, and political impact.

Governments and international organizations are conspicuously absent and inert. It is as if they believe that technological progress is an end in itself. They care little about unlawful behavior taking place on the Internet. They leave it a "free-for-all," considering the risks to be inevitable, as if progress entails certain unavoidable and uncontrollable ills.

Major technology companies run campaigns with silly slogans, such as "tech for good" or "tech for safety." Google, Twitter, and Facebook all have sonorous mission statements such as "don't be evil," "help increase the collective health, openness, and civility of public conversation," and "give people the power to build community and bring the world closer together." Of course, it would be naive to expect these companies to care about the social problems created by innovation and the technological revolution.

Public opinion thus remains disoriented and, to a certain extent, manipulated by voices that are neither competent nor neutral. Some exaggerate the risks, while others promote tolerance and the take-up of technologies by downplaying the risks. The public is ready to accept the wrong ideas and to conform.

The increasing automation of processes and services that are critical to the seamless functioning of the economy and the organization of society leads to a centralization of decision-making with regard to control of the cyber-world. The problem is political in nature: the democratic control of decision-making centers for the rational and safe use of infrastructure and services.

6.4.2.3 Technological Dependency

The prevalence of a technology that solves a host of practical problems and makes life more comfortable means that certain skills that we developed to solve such problems fade away. For example, very few people today know how to light a fire using friction, a skill prehistoric people have fully mastered. Nor do people today know how to survive in the wild or build a hut for protection. In the future, children may no longer learn the multiplication table, which was fundamental in children's mathematical education up to and including the twentieth century. It is reasonable that the question of over-dependence on technology would be raised. Which basic skills/knowledge must we absolutely maintain? The question is now being raised even more fervently, since technology not only solves individual problems, but also provides comprehensive solutions that imply a different way of life where it relieves us of the burden of managing decisions by offering a plethora of services.

In order to explain the risks involved, certain people refer to the fable of the "boiling frog" [5] placed in a pot of water. If we were to suddenly raise the water temperature, the frog would jump out of the pot. However, if we were to gradually raise the temperature, the frog would initially feel pleasant and would remain in the pot until it died.

Currently, users of social media are willing to provide, in exchange for services of dubious quality, their personal data, which is a valuable basis for controlling public opinion. Each person's personal preferences and information certainly have no commercial value for them. However, the result of intensive analysis of very large datasets is crucial for the predictability and control not only of resources and systems, but also of markets and behaviors. Perhaps we do not understand the importance of this information since it remains secret and solely at the disposal of those in power and those who pay a handsome price for it.

I discern two threats to individual liberties.

1. One is the violation of privacy in order to allegedly protect societies from infringing behavior, such as terrorism or crime. There are many plans to increase the monitoring and control of individuals by developing appropriate technological solutions. I would mention, for example, *reputation systems* that allow rating each other in online communities in order to build trust through reputation [6].

   Needless to say, it is necessary to have a regulatory framework for the use of such tools that violate privacy and can be used to stigmatize or exclude citizens through arbitrary processes and criteria.
2. The other threat comes from the increasing use of autonomous services and systems in the name of efficiency. This vision is being promoted through the Internet of Things. It envisages automation of critical resource and infrastructure management systems without human intervention. Decision-making criteria may be so complex that they exceed human possibilities of understanding and control. We may thus reach a system of techno-tyranny, all the more because automation enables the concentration of decision-making at ever fewer centers.

   Therefore, the real risk is that computer knowledge will be used uncontrollably to make decisions, replacing people in critical processes.

The proper and rational use of artificial intelligence and autonomous systems depends on two factors.

1. The first factor is the possibility of assessing, based on objective criteria, whether we can trust the knowledge generated by computers. This is currently the subject of research which, one is to hope, will be able to give us the necessary results.

We are trying to develop "explainable" artificial intelligence using techniques that allow us to understand and, to some extent, control the behavior of systems.

I believe that we need a new method of assessing knowledge, bridging the gap between scientific knowledge and the empirical knowledge of artificial intelligence.

2. The second factor is the vigilance and sense of political responsibility of society as a whole. When computers use knowledge to make critical decisions, we need to be sure that this knowledge is safe and neutral. The safety of autonomous systems must be certified by independent authorities and not be left to those who actually develop them. Here, the existence of a legislative and regulatory framework and the role of institutions in risk management, which will be discussed in subsequent chapters (see Sect. 7.3.2), could come into play.

References

1. https://en.wikipedia.org/wiki/Thinking,_Fast_and_Slow
2. https://en.wikipedia.org/wiki/History_of_artificial_intelligence
3. https://en.wikipedia.org/wiki/Correlation_does_not_imply_causation
4. https://en.wikipedia.org/wiki/Technological_singularity
5. https://en.wikipedia.org/wiki/Boiling_frog
6. https://en.wikipedia.org/wiki/Reputation_system

Part III
Consciousness and Society

We seek to analyze the functioning of individual consciousness and to discuss its impact on societal organization, in the light of the previous gnoseological perspective.

- We explain the functioning of consciousness as an autonomous system that manages short- and long-term goals based on value criteria and accumulated knowledge.
- We discuss how values are shaped in society and how the composition of each individual's subjective experience takes on an objective dimension, which can be studied as a social phenomenon.
- We present an analysis of the role of institutions in shaping and maintaining a common scale of values for strengthening social cohesion, as well as a discussion on the principles of democracy.

Chapter 7
Consciousness

7.1 Our "Big Bang"—Language and Consciousness

7.1.1 Language and the Boundaries of Knowledge

Of all the seemingly insurmountable problems we face, one looms the largest and most important. How did cognition and language appear through evolution? How was consciousness formed? Questions concerning cosmogony or the evolution of the species appear trivial in comparison to the question, which sadly has been hitherto subject to fragmented research in various disciplines (biology, psychology, informatics, linguistics, anthropology, etc.). How are abstract concepts created through experiences? How do we comprehend by linking phenomena to analogies and metaphors? How are the boundaries of knowledge and understanding of the world demarcated? How do we ascribe meaning to symbols? How do we create? How does awareness function as a system that combines wishes, motives, and will? How do we choose goals and act on them?

Language defines the coordinates of cognition: being able to mean what you say and to clearly state what you mean. What you feel is more than what you can express. What occurs is more than what you can describe. Only what can be placed in the semantic relationships of language can be understood. By language, I mean not just natural language, but also internal and ineffable language, as well as any manner of structure that can be interpreted and carry information.

Important research programs are focused on studying the functions of the human brain. I do not question the importance of such research, which will teach us much that is useful about the functions of this wondrous "processor." However, I do not think that, due to their very nature, they are able to answer the questions at issue. Mental phenomena must be studied on a different scale and cannot solely involve analyzing the signals and neural circuits of the brain.

Many will claim that these problems can only be the subject of philosophical inquiry, and that a strict scientific approach is impossible or even misplaced. I would

J. Sifakis, *Understanding and Changing the World*,
https://doi.org/10.1007/978-981-19-1932-9_7

counter that such an assertion must be supported by arguments as to what exactly a scientific approach is and what its limits are. However, this is part of the questions we have raised.

In the past, problems, which were the subject of philosophical inquiry, entered the sphere of scientific analysis. Mental phenomena complement physical phenomena. They are the other side of reality. Attempting to understand them through scientific methods is a worthwhile, imperative endeavor. Such an undertaking requires broad interdisciplinary cooperation, and I believe that the contribution of disciplines, such as informatics, biology, and medicine, could be decisive.

In the future, I would like to see the same generous funds and mobilization of large research teams dedicated to exploring space being used to set up programs to explore the mind as a tool, which on the one hand, ascribes meaning to the world and, on the other hand, is the starting point for decisions on action and creativity. Perhaps then we might be able one day to penetrate the mystery of the "Big Bang of consciousness," and the theories of physical cosmology might be complemented by other, parallel ones that would illuminate our unexplored selves and bring us slightly higher on the wonderful scale of self-knowledge.

To what extent can we formalize natural languages in order to be able to delve deeper into their concepts and relations? This remains an open problem despite advances in mathematics and logic. Formalization requires a set of core independent concepts to which all others refer. When concepts are not independent, they should be linked through semantic relationships, as is the case in formalized languages. What makes formalizing natural languages very difficult is the lack of adequate semantic models, as explained in Sect. 6.1.2.

One major advantage of humans is that natural languages are discrete. Phrases consist of words, and words consist of syllables.

How would we understand the world if language was not discrete, but continuous? What if we talked in whistles, like birds? With no consonants or vowels? I am not referring to well-known whistling languages that simply translate discrete natural languages. I am imagining languages where every concept or phrase would be a whistle, which, in my opinion, would severely restrict their expressiveness. This would depend on our acoustic capabilities—that is, the frequency interval of the acoustic spectrum that we could perceive.

If that were the case, I cannot imagine how we would define mathematics and algorithms. Could an algorithm be a transformation from one continuous signal to another? I believe this passage from the continuous to the discrete is integrally linked to the development of conscious thinking.

7.1.2 The Mind–Brain Relationship

Another interesting question is the mind–brain relationship. Can all mental processes be understood by exclusively studying the brain? Let me express my doubts by proposing an analogy. Let us suppose that your computer evaluates a mathematical function and that an engineer can observe the state changes of all its circuits during the computation. Is it possible, from observation alone, to determine the function whose code the computer executed? Without going into technical arguments, the answer is that this is an extremely difficult problem. I believe that we will not be able to understand mental functions by exclusively studying the brain as a physical and information processing system (see also the discussion on emergent properties in Sect. 4.1.4).

We currently have techniques at our disposal to study how concepts are represented in the brain as a pattern of electrical signals. One interesting question is whether the way in which data are represented and generally organized is the same for all individuals, and especially whether it depends on the language we speak.

Recent thought transmission experiments using computers (Brain Computer Interaction) tend to demonstrate that concepts have a canonical representation in the brain. In other words, the same concept, regardless of language, is represented through the same combination of signals and states (patterns). If this is the case, it should raise concerns for those who believe that we are the result of random evolution. This would mean that, although we do not speak the same languages, there are common origins in terms of the representation of concepts. If it is experimentally proven that the representation of a concept such as "door," "window"—or much more so "freedom"—has common characteristics regardless of language, this will be a very interesting result. It may even shed light on the problem of the existence of language families.

The theory of the existence of a Proto-Indo-European language that developed in a certain region and then spread with differentiations across almost all of Eurasia does not seem so convincing to me. The solution to the mystery should be sought elsewhere. Research into the representation of concepts in the brain may help us to reach some interesting conclusions.

7.2 The Mind as a Computing System

I have written this text as a computer scientist and engineer, relying particularly on models and ideas that I have developed for autonomous systems that I have been studying in recent years. As a non-expert, I have no claim to the "scientific veracity" of my views. It is rather a conjecture into how the human mind could function as an information system based on my knowledge and what I have understood by reading relevant texts from psychologists, neuroscientists, and philosophers. I would like to stress that these texts contain a fairly wide range of views and are characterized by an absence of organization of concepts and the relationships between them. As a reader, I found very little that was actually possible to digest.

What I tried to do is to imagine an architecture where the various elements of consciousness function in a technically feasible way, insisting more on relationships between concepts that determine the flow of information and knowledge, without this meaning anything about the model's accuracy regarding what might actually be occurring.

I once wrote that scientific knowledge is nothing more than a set of myths, which have been tested in practice with systematic experimentation and validation. Such myths make airplanes fly, keep bridges and buildings standing, and illuminate continents when the night falls. However, they are but a poor approximation of physical reality, to the extent that we can observe and understand it.

7.2.1 Value Systems and Regulatory Frameworks

Based on which mechanisms and criteria does the mind exercise freedom of choice, if, of course, it can actually choose? If so, how does it choose between many diverse needs, both material and intellectual, when it has limited resources?

One simple way to understand these mechanisms is by defining a common benchmark.

Let us consider that each individual has a *value system* using *scales of values* to determine for each action, the units of value required for it to be carried out and, possibly, the units of value it produces. Such an idea has been successfully implemented in many fields of knowledge. It is the idea of the "invisible hand" that Adam Smith imagined for the economy. It is the idea of the *resource* for computing systems, which I explained in Sect. 5.2.2. The laws of physics can also be regarded as a *value game*. If I want to move and go from one place to another, I need resources, energy in this case, which depends on my mass and the relationship between the coordinates of the positions.

Of course, it might seem strange to use numerical values, considering that any kind of human action has a "value balance." For example, how would I measure the intensity of the feeling of hunger, which I can "zero" with a meal that costs 15 euros? However, if we wish to find a functional scheme that correlates the various criteria entering the consciousness game, we have to accept a "common denominator" and "equivalence relationships" in order to compare needs against each other and match them with the resources needed to satisfy them.

I will also consider that any social organization in a place and time is based on a *common value system* that emerges as the "synthesis" of the individual value systems of its elements. Similarly, this system uses *common scales of values* for evaluating the actions of individuals, enabling us to understand how societies and individuals function as dynamic systems, more or less coherently. This is achieved thanks to the existence of a common framework for evaluating actions, encouraging and rewarding beneficial actions, while discouraging and penalizing actions that harm society as a whole. Hence, social cohesion and the synergy of individuals to achieve common goals are reinforced.

The values of common scales and their nature change according to the field of action, i.e., economic, political, legal, educational, military, gnoseological, moral, religious, and esthetic. We sometimes relate values to *principles* because they indirectly define what is important or even provide a basis for comparison in order to set priorities between our actions and appropriate incentives to implement them.

For example, the economic value system concerns monetary values, which are regulated in order to help the orderly production/exchange/use of goods and services. The incentive in this case is to generate profit.

Political values encourage good governance by making timely and key decisions for societal prosperity and safety. Legal values are used to administer justice by preventing the violation of laws through penalization. Educational values, such as excellence, and enthusiasm for learning and creativity, are intended to support the transmission of knowledge and to prepare citizens capable of contributing to and integrating into society. Military values foster a spirit of readiness and self-sacrifice to safeguard a country's security.

Furthermore, moral values define what is good in our behavior, while gnoseological values promote sound knowledge and information. Religious values aim to strengthen relations with the divine through a set of rules and practices. Finally, the scale of esthetic values sets criteria for what is beautiful or pleasant, such as a work of art, an environment, or an experience.

The above list is not exhaustive and is provided simply in order to explain the concepts introduced.

I will assume that we use scales of values to correlate the values that each action consumes and generates. Positive value means profit, recognition, satisfaction, for example, when I pass a test, or when I earn money. Negative value means loss of money, a penalty or shame, for example, when I am arrested because I violated road traffic rules or when I am seen mistreating an animal.

A specific category of actions with a negative value are prohibitive actions with a dissuasive cost for the person who commits them, conferring upon such actions a deterrent effect. These include criminal offences, such as killing or robbing someone, which are punishable by law, as well as actions, which, albeit not punishable, involve excessive risks, such as spending recklessly and racking up debt.

Scales of value also define the *mandatory* actions that we should reasonably be required to take because doing nothing would entail prohibitive value costs. Thus, you have a legal and moral obligation to help someone in danger, to respect road traffic rules, etc.

Note that prohibition and obligation are not independent concepts: a prohibition could be expressed as an "obligation not to." We can therefore say that "it is forbidden not to comply with road traffic rules." However, in practice we use both concepts for the sake of clarity.

Actions that are not mandatory or prohibited are *optional*. They involve a relationship of reciprocity. The criterion for their execution is whether their performance is considered to be advantageous. Optional actions define the framework for individual freedom where public conscience is exercised with a view toward the common good and societal prosperity.

It should be noted here—and will be made even clearer later—that, given the heterogeneity of resources (time, money, physical and intangible resources) and quality criteria, it is not simple for one to decide what is in one's interest and what is not. In taking a certain action, I may use money to gain time and/or social recognition that cannot be quantified.

Furthermore, an action can also affect more than one field. An economic action can have consequences that are not only economic but also social and moral. On the other hand, a morally motivated action, for example, charity, can have economic effects.

Although the values of the various fields are not comparable, when we make a choice, whether consciously or automatically, we are using an equivalence relationship. For example, when you decide to pay 200 euros in order to buy a pair of shoes, you take into account your financial capability, as well as other esthetic and social values that do not arise from your need for footwear.

Often, one way to promote values that concern non-material goods, such as road safety or patriotism, is to encourage practices or to apply regulatory frameworks that lay down mandatory or prohibitive *rules*. Thus, road traffic rules promote road safety and are part of the statutory regulatory framework. Right-hand priority rules or speed limits restrict the behavior of drivers in order to enhance safety. Therefore, the application of a traffic rules system allows for something that cannot be practically achieved by abstractly explaining to the general public that road safety is an important value. That is why social organization around a value system is achieved through the application of regulatory frameworks. There are countless examples in every field.

If virtue is considered to be a supreme moral value, ethical rules help you to be virtuous either by preventing you from acting immorally or by indicating how you can be virtuous in certain circumstances, for example, by encouraging charity or mutual aid.

In education, there are rules that encourage and recognize excellence, for example, students who excel are valedictorians at graduation ceremonies, are awarded scholarships or enjoy other advantages. The above demonstrate the importance of regulatory frameworks for social organization, which we will further discuss in Chap. 8.

In order to study the behavior of individuals, I will assume that each person has an individual value system that largely reflects the values of a common value system, especially those that we cannot disregard without cost, for example, financial or legal. The individual value system also includes rules regarding values that characterize actions of a personal nature, such as in the field of ethics, religion, or esthetics.

7.2.2 The Components of Consciousness

It is not easy to define consciousness, as the concept is quite complex and has not been fully explored. I will begin with a simple definition, which will be enriched in this chapter, by saying that consciousness is the ability of humans to understand the world and act in response to internal and external stimuli. Consciousness manifests itself as an interactive game between the mind and its environment. On the one hand,

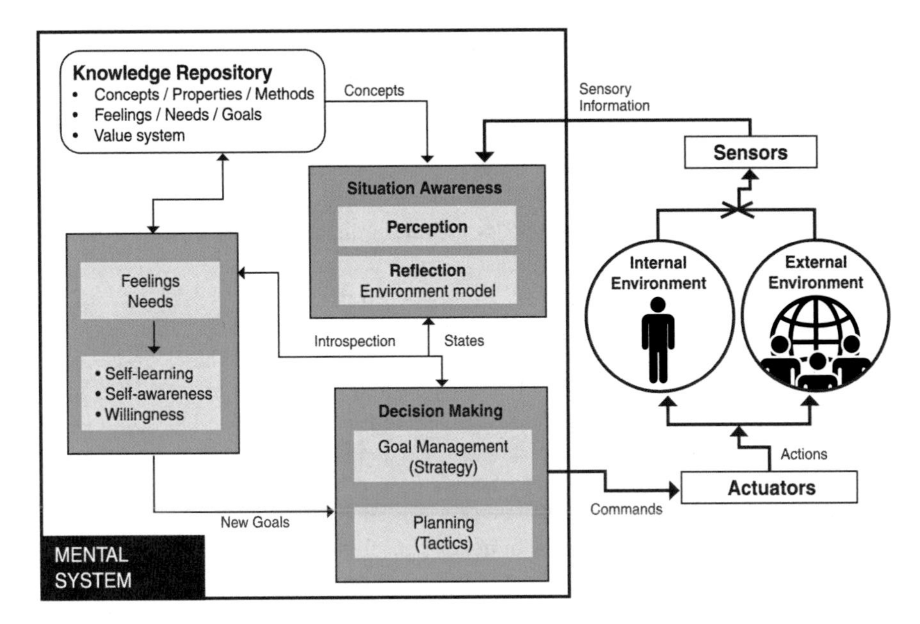

Fig. 7.1 Architecture of a mental system with its key functions

the mind plans and carries out actions to meet the physical and non-physical needs of the individual. On the other hand, the environment determines the economic, social, and physical conditions in which individual freedom is exercised.

In this game, one's individual scale of values plays a central role in exercising freedom of choice and achieving one's goals.

I am not seeking to explain exactly what consciousness is and how it actually works. My aim is to define, as simply as possible, the architecture of an information system, which I call a *mental system* and whose behavior approximates that of conscious thinking. I use the term "approximates" because I focus on decision-making issues and I ignore others, perhaps more important, that concern self-learning and self-awareness, as well as cooperation between the conscious and the subconscious.

I assume that the mental system is an autonomous system (Fig. 7.1), similar to the one described in Sect. 6.3.2, suitably enriched to take into account the particularities of humans.

The mental system interacts with the external environment by obtaining sensory information and performing actions. In the case of humans, the external environment is the entirety of external phenomena, both physical and social. The internal environment is the body with the senses and the motion and speech organs. By using the functions of perception and reflection, the mind builds a model of the environment, as I have already explained.

One particular feature of human beings is the ability to self-learn. Humans can create new goals to meet different kinds of physical and non-physical needs. This admirable adaptation mechanism is based on a capability for introspection, which interprets the states of the environment model and produces *feelings*, which I will discuss immediately below. For the moment, I will say that feelings are mental states, which are accompanied by emotions and are directly linked to needs. When the intensity of a feeling exceeds a threshold, the will to satisfy the need arises by setting the appropriate goals, to be managed by the decision-making function. Goals are determined by taking account of the individual scale of values, the state of the environment and accumulated knowledge regarding the environment, expressed in the form of general rules.

I consider that this architecture of the mental system, which I will explain in further detail, consists of three components dealing with: (1) understanding and building a model of the environment with which we interact; (2) the emergence of needs and the will to satisfy them by setting the relevant goals; and (3) decision-making concerning the management of goals, and their planning and realization.

In conclusion, I would point out that Plato also distinguishes three key parts in the human soul, which are not unrelated to the functions of the model: the appetitive part (*epithumetikon*), the spirited part (*thumoeides*), and the rational part (*logistikon*). All the desires of a person, as expressed in his or her interaction with the environment, emanate from the appetitive part. The will to satisfy needs by taking account of a set of values emanates from the spirited part. Finally, the rational part concerns the coordination of the other parts and the management of knowledge.

7.2.3 *The Mental System*

7.2.3.1 Feelings and Needs

Through the process of introspection, the mind creates mental states that describe the intensity of *feelings*. I will not attempt to define the concept precisely, as there is no agreement among experts. I will simply say that a feeling is a subjective conscious experience, combining mental conditions and psychosomatic reactions.

As mental states, feelings have intensity and a "sign." Generally, a feeling is pleasing when it satisfies criteria concerning the entire spectrum of quality of life, from comfort to survival. The criteria are subjective but are influenced by objective factors. Therefore, for some reason I may be feeling cold, even if the temperature is not low. The intensity of feelings is different for different people under the same environment conditions.

Feelings are linked to the satisfaction of needs: biological, safety, social, self-esteem and self-fulfillment, intellectual, etc.

Biological needs are expressed through feelings of thirst, hunger, sleep, fatigue, and others concerning the satisfaction of physical needs. Safety needs concern the stability of work relationships, adequacy of financial resources, living conditions, social security coverage, etc. The related feelings, such as fear or anxiety, stem from our assessment that we are facing a risk, based on the state of the environment model and our knowledge from previous experiences.

Satisfying social needs manifests itself through positive feelings stemming from the existence of stable and substantive esthetic, friendly, and familial relations. Failure to satisfy such needs entails feelings of rejection and abandonment.

Self-esteem and self-fulfillment needs are expressed through positive feelings, such as self-confidence, professional recognition, and feelings of excellence due to a position in a hierarchy.

Finally, intellectual needs are grounded in non-material values and are expressed through feelings such as compassion, solidarity, temperance, and honesty.

Experts suggest detailed classifications of feelings and the related emotions. I am not competent to speak on the matter. The above list is not exhaustive and is provided simply for argument's sake.

I highlighted the distinction between positive and negative feelings. Another interesting distinction is between *passive* and *active* feelings.

Passive feelings are those that do not urge a response. They stem from a situation that we are experiencing and that does not give rise to other feelings. Such feelings are fear, sadness, shame, and pain.

Fear, for example, stems from the assessment that we are in a dangerous situation. This may be objectively justified by a reasonable analysis of the environment state. If the fear we feel is exaggerated, then we are talking about a phobia. When travelling by airplane, there is always a minimal risk, which is assessed according to technical criteria. The behavior of those on board, reacting as if there is a high likelihood that the worst will come to pass at any moment, can be considered a phobia.

The opposite feeling, fearlessness, is active and means a rash underestimation of risk. Examples include participating in an extreme sport without taking the necessary precautions, or indiscriminately spending money and reducing our financial resources beyond the threshold necessary to survive.

Active feelings suggest situations involving alertness or stimulation of mind, such as enthusiasm, passion, interest, ecstasy, and responsibility.

For example, a feeling of responsibility means constant vigilance to weigh the pros and cons of our actions and to take their possible consequences into account. Being responsible does not just mean doing the right thing according to certain criteria, but also acting with all the care necessary to achieve the best result. It is easy to prohibit children from doing something they should not do, but it is more responsible to convince them to do the right thing through dialog.

An irresponsible person will take a cavalier attitude of over-optimism regarding the outcome of their actions, perhaps taking risks that could lead to dangerous situations. A responsible person will endeavor to weigh the risk by using their knowledge and objective criteria as much as possible. They can thus justify their actions as a result of an analysis that takes various criteria into account, ranking them hierarchically.

A responsible person has a reasonable attitude toward risks. One interesting case is heroism, when a dutiful person overcomes fear. Heroes defy fear not because they do not properly weigh the risk, but because they are driven by a sense of duty. They consciously carry out a dangerous action, considering the defense of intellectual values superior to their own personal safety.

How do feelings relate to needs? Having assumed that every feeling relates to a need, I believe a need arises when the intensity of a related feeling exceeds a certain threshold.

I will represent the intensity of feelings with variables defined in an interval of values between a lower and an upper bound. Conventionally, positive values express satisfaction, negative values express the opposite, while 0 is an indifferent state. We have pleasant feelings (positive value), unpleasant feelings (negative value), or even indifferent feelings (zero value) by assessing the states of the environment. Thus, I may feel indifferent toward sports news, the fact that whales are being decimated en masse or that climate change is looming over us.

For example, I can use a scale between +5 and −5 to express the intensity of the feeling of hunger. Conventionally, −5 means I am absolutely sated, +5 means I have an absolute need for food, and 0 reflects an indifferent state. Of course, every person has his or her own personal tolerances.

The intensity of feelings on an individual scale of values plays an important role in the choice of actions and goals.

I believe that for every feeling there is an optimal value which I will call a *point of equilibrium*. It is the most painless state for negative feelings and the most pleasant state for positive feelings. Thus, I can imagine that the mind chooses the actions carried out, so as to be close to the point of equilibrium for all feelings. Such a situation is difficult to achieve. This is because, when I try to change the intensity of a feeling, for example, to restore the feeling of hunger back to 0, the point of equilibrium, I may disrupt the intensity (state value) of other feelings.

I have said that a need arises when the value of a feeling exceeds certain limits on the value scale. If, for example, the intensity of the feeling of hunger approaches critical values, then the system inquires whether and how this need can be satisfied by choosing appropriate goals. In other words, a goal concerns the attainability of states of the model that satisfy a condition—in this case, access to food. The correspondence between needs and goals is knowledge developed through the self-learning function based on experiences. That is, I know from experiences how and where can I find food if I am feeling hungry. Goals are selected through the decision-making functions of the mental system, as explained in what follows.

7.2.3.2 States and Actions of the Mental System

The mental system is a system with states and actions, such as the one we presented in Sect. 5.2.2. It has a very large set of states defined by the values of three types of variables.

1. External variables describe the state of the external environment.

   For the physical environment, variables include those that determine location and time, weather and environmental conditions such as temperature, pollution, and noise.

The economic environment is characterized by a set of variables that have a particular impact on the assessment of one's personal financial situation in order to take actions of an economic nature. These concern the income, debts, taxes that we pay, but also other, more general parameters, such as the level of interest rates or property prices.

Other variables are those that affect decisions on all types of social services, such as education, health, and protection, for example, the cost of medicine and hospitalization, and the level of social benefits.

2. The variables of the internal environment characterize our body's state according to physiological parameters such as temperature, heart rate, and blood pressure.
3. Finally, the model also possesses a variable for every type of feeling that describes its intensity using values of the respective value system.

Furthermore, the system has two types of actions:

1. Actions *controllable* by the mental system, which may decide whether to carry them out when their preconditions are met, for example, withdrawing money, travelling, and shopping.
2. Actions *non-controllable* by the mental system, carried out in its environment by other persons or with natural causes. Actions by external factors, such as the expiry of a deadline or a change in the weather, can alter the state of the mental system regardless of our intention. There are also internal non-controllable actions, which are automatically carried out by internal mechanisms and lead, for example, to feelings of hunger, thirst, etc.

When we perform a controllable action from a state, the mental system transitions to a new state, changing the state variables according to the amount of values that the action creates or consumes, as defined by the individual value scale. This action, as we have said, cannot ignore the common scale of values, particularly with regard to changes in the variables of the external environment.

As in Sect. 5.2.2, an action is defined by the *preconditions* for its execution at a state of the model and by its *result* when performed.

1. The preconditions for an action to be executed at a state are qualitative or quantitative requirements that the state must satisfy. The quantitative requirements relate to the availability of a quantity of physical resources and/or intangible goods measured on the applicable value scale. Resources are committed and/or consumed to render the action feasible. For example, a precondition to go shopping is to have the necessary funds available. A precondition to dare to go skydiving is to be in the right physical and mental state to overcome the feeling of fear. Qualitative preconditions are assertions about the system state that either hold true or not. For example, in order to read an e-mail message, you must have access to a network and an electronic device.
2. The result of the execution of an action is to move from the present state to a new state by changing values of state variables, in particular by consuming and creating values. For example, when cooking baked pasta, I use resources, namely basic foodstuffs and energy, and I produce, for example, 10 portions of baked

pasta. When I sleep, I consume "active" time and proportionately increase restfulness.

Not only controllable actions but also non-controllable actions can affect our mental state. This is because, through our senses, information regarding a change, for example, in room temperature, in bank interest rates, and in a building's plot ratio can affect our feelings, just like the satisfaction or non-satisfaction of a need for heat results in optimism or frustration.

By extension, an action stemming from altruistic feelings can also change external variables. Helping an injured or disadvantaged person is an action that increases my self-esteem, but also involves interaction and therefore changes to the external world.

It is interesting to understand how material and non-material values are linked when we perform an action, such as "having lunch at a restaurant." There are preconditions such as "being at a restaurant," "having the necessary funds," and "the feeling of hunger exceeds a certain value." Immediately after the meal has ended, I will still be at the restaurant, but the funds available will be fewer, depending on the amount spent on the meal, and I probably would be quite sated (satisfaction of appetitive needs). Here I am using the value scale for the feeling of hunger. We are not interested in exactly how it is determined and how it is graded, which may vary from person to person. However, such a concept is necessary in order to study the effects of hunger on the behavior of individuals and their environment.

7.2.3.3 Selection of Actions—Conflicts

We already discussed conflicting actions in Sect. 5.2.2: when two or more actions from one state are possible and realizing one can cancel out the preconditions for realizing the other. There are many everyday examples of conflict situations, which the mind must resolve in the best possible way between the controllable actions of the model from a given state. First, the choice is made between the actions whose preconditions are valid. We know that every action can consume and release values. Therefore, a simple selection criterion is to choose an action whose balance of values is the "most positive" one. For example, I would prefer to use existing financial resources in an investment with a high yield rather than buy an apartment or travel around the world. Of course, the problem of comparing dissimilar values arises, but it is not as theoretical as it seems, as we address it every day. Equivalences, even if non-consciously, are applied in order to compare different value scales.

Therefore, you decide to perform charity because the moral satisfaction that arises is more important than the financial cost it entails. The values of material goods are transformed through the market into monetary values—but there are also immaterial goods, such as freedom, justice, and knowledge, which should be estimated on a value scale in order to resolve conflict situations.

Individuals and societies must choose between non-comparable values. How much can one exchange one's liberties for material goods and comfort of life? There are many such questions, and the answers we give them at a given time depend on the way values are hierarchically ranked, as shaped less in the form of written rules and more in our conscience.

Another criterion for choosing an action is the criticality of a situation. For example, lack of a resource can have negative effects, such as an impasse when the intensity of a feeling is far from its point of equilibrium. Thus, if my funds are dramatically reduced, then the choice of "emigrating to work abroad" can become a priority over other choices, even if it entails a high emotional cost. Similarly, if I am very tired or hungry, the actions that will bring me back to the point of equilibrium take priority over others that may have a more positive balance of values.

It is worth noting here that the results of actions do not always have the same sign for both the individual and society as a whole. One characteristic case is that of a criminal, whose action may be harmful to others, while its outcome is positive for him or her.

The behavior of an upstanding citizen is such that they do not take actions that are harmful to others, even when they suffer harm themselves. There is, of course, also the case of the fools, who may indiscriminately take actions that harm both themselves and others.

This simple model that views the conscious execution of an action as the result of a decision, weighing the pros and cons of prerequisites and results, enables us to understand how differences in people's value systems result in differentiation in their behaviors.

This differentiation is what makes a person stand out as selfless, selfish, gourmet, heroic, rude, etc. A selfless person will take an action for the common good, even if they stand to take a loss. A selfish person would not accept any action that would entail a personal cost, no matter how favorable it would be for others. A gourmet may place culinary values higher than monetary values in comparison to the average person. A hero would take action for the common good even if they entail a prohibitive cost. A rude person would act in a way that does not respect established rules of behavior toward others.

In conclusion, I would say that the mental system aims at the harmonious operation and oversight of the whole, while maintaining a balance between satisfying material needs and value requirements. This cooperation/struggle between desire and spirit or mood, under the arbitration of informed thinking, results in highlighting short- or long-term goals.

In what follows, I will try to explain how the elements of consciousness work together to achieve goals.

7.3 Goal Management and Planning

7.3.1 Strategies and Tactics for Achieving Goals

When there is a need that cannot be satisfied through immediate action by the current state of the mental system, the mind must identify possible goals, i.e., attainable model states through which it can be satisfied. We already said in Sect. 6.3.2 that there are positive and negative goals. Satisfying a need can be achieved through different positive goals to different degrees. However, every positive goal also results in other negative ones. Negative goals are linked to the cost of *risk*, as they concern avoiding dangerous situations entailing considerable value cost. Risk depends on the possibility of finding ourselves in such situations due to various uncertainties.

For example, an economic action involves transactions with other persons who may not be solvent or who are affected by political and social changes. When we decide to partake in a dangerous sport we enjoy, we take the relevant risk into account when calculating the cost. This also applies when we travel or generally change our state within the environment. There is also the risk of facing legal consequences when we act unlawfully. Finally, there is the risk that arises from satisfying needs when the action may endanger our physical integrity.

Therefore, when assessing the cost of achieving a positive goal, we must also incorporate the risk of the associated negative goals. This is an extremely complex problem, because a tactic is not a simple sequence of actions, but a set of sequences of actions depending on the reactions of the environment. Thus, the cost estimate must be graduated taking into account the "worst case scenario" and the resources available.

7.3.1.1 Strategies

In a given state of the mental system, a set of needs and corresponding goals emerge in the mental system, which the mind initially ranks depending on their importance and criticality. As is the case with actions, a goal becomes critical when failure to satisfy the need in time can lead us to catastrophic situations with huge value costs, whether material (money, health, and integrity) or immaterial (deprivation of liberty, moral cost). Thus, the mind tries to achieve critical goals in order of priority, planning appropriate tactics and mobilizing the necessary resources.

Non-critical goals may also be classified according to the cost of not achieving them in time. The mind makes comparisons between similar goals, for example, to go to a restaurant or to the movies on a given night, both needs arising from the desire to "blow steam off," taking into account all the knowledge relevant to their achievement. It will choose from a variety of economic and technical criteria (how easy it is to visit each locale from one's home) and personal preferences and memories from any past experiences. When we have satisfied one of the two goals, on the way back home we make a conscious or even automatic assessment of whether the choice made was good (whether it had the results anticipated). If not, then the self-learning function will change certain model parameters that will affect the relevant choices in subsequent judgments.

The hierarchization of goals into short- and long-term goals is taken into account together with their prioritization. Short-term goals are those that we manage every day and have a sense of urgency. There are also exceptional short-term goals that are usually given higher priority.

Long-term goals must be achieved over time and, to this end, the mind must carry out a more thorough analysis in order to define a strategy, taking into account the resources available and external factors. Their management requires special diligence and perseverance because we often tend to be diverted by short-term goals, even if they are not important.

We can consider the development of strategies to manage goals as a solution to an optimization problem, taking into account constraints related to the resources available. We will choose a set of goals whose timely achievement is possible using the resources available and which are expected to best satisfy our needs. One inherent difficulty in this process is the lack of predictability of dynamic changes in the environment. New goals may emerge or existing ones may need to be modified. The more predictable the environment, the more successful our goal management can be. This is perfectly understandable in the field of economics, where stability and predictability of trade allow for proper resource management. Lack of predictability implies a permanent commitment of resources to respond to emergencies. That is why social and political stability are important factors for economic growth and prosperity.

7.3.1.2 Tactics

When a goal is selected, the mind calculates a tactic to achieve it. The tactic consists of planning sequences of controllable actions that may lead to achieving the goal depending on uncontrollable environment reactions. Here lies the complexity of calculating tactics. A tactic is not a sequence of actions, but a set of action sequences that can be represented by a tree structure. When we perform an action that changes the state of the environment (node of the tree), the next action is chosen in response to the reactions of the environment (branches of the tree from the node). When the environment offers a large number of reactions—as is often the case in practice—it is impossible to make a static estimate of tactics to achieve a goal from a given state. A tactic is dynamically calculated over a time horizon and updated from time to time until the goal has been achieved.

Thus, calculating tactics appears to be a game that we play against the environment, both external and internal. The difference, though, is that in this game the rules are not clearly defined, particularly regarding the intensity and the timing in the exchange of "blows" between opponents.

Identifying successful tactics is not easy and requires not only knowledge, but also proper knowledge management in order to both assess situations and find solutions that keep the goal accessible.

It is sometimes greatly tempting to take an action which brings us very close to achieving the goal, but at the same time places us in a situation where it is impossible to go further—and so we definitely lose the game. Examples could be making an investment by taking out a loan that I must start paying off before I reap the return on my investment or running a race and exerting all my energy to remain first at the outset, which I cannot keep up all the way to the finish line.

I must point out that, in practical terms, the effectiveness of the decision-making mechanism depends on the quantity and quality of knowledge, but above all on the ability of the mind to effectively combine knowledge. This is what I call meta-knowledge or wisdom: a person can have a wealth of knowledge to answer basic questions but be incapable of combining them creatively to achieve a goal. There are well-read persons who are "living encyclopedias" yet are inadequate in managing and achieving goals—and there are people with a moderate education with outstanding judgment and inventiveness in difficult situations.

7.3.2 Safety and Liberty

"Those who would give up essential liberty to purchase a little temporary safety deserve neither liberty nor safety," said Benjamin Franklin. I agree with this phrase, which has a lot of truth in it and is a fine defense of fundamental liberties.

We must not ignore the fact that liberty and safety are often opposing concepts. When one grows, the other diminishes. The issue is which liberties we are prepared to sacrifice and to what extent in exchange for our safety. It is up to each person to strike the right balance between liberty and safety, based on their values and the estimated degree of risk.

Each person must exploit his or her degrees of liberty to achieve a goal effectively without jeopardizing his or her safety.

We have already discussed critical goals, which if not achieved, result in catastrophic situations. However, we face the daily risk of such situations if we do not plan ahead so that our actions are safe.

An action is not just safe if its result does not directly lead to a catastrophic state. Additionally, after its execution, it should not be possible for the environment to lead us, through non-controllable actions, to a catastrophic state without us being able to prevent it. For example, if we leave the house without locking the door, we are creating a potentially unsafe situation, in the sense that a person can break in and burglarize our home (catastrophic state).

Technically speaking, safety is described by conditions on variables of the mental system that characterize safe states. Examples would include an annual income of over 10,000 euros and a body temperature of under 40 °C. Non-satisfaction of a safety condition can lead certain actions to deadlocks. If a deadlock is permanent, this entails a significant decrease in the degrees of liberty in the management of goals.

One special case of safety conditions are those subject to time constraints. In theory, if we took no initiative for action, our state would not change were it not for the factor of time. In a sense, time acts as our "adversary." Over the course of time, even during a period of complete inertia, we have to eat and drink, pay our rent, go to sleep. Thus, there are conditions determined by critical deadlines, which if not met, can lead to an adverse result.

Complying with safety conditions generally results in reducing degrees of liberty and, therefore, effectiveness in achieving goals. If I wish to advance too quickly, I might disregard or fail to properly assess certain risks. For example, I might not wait at a street crossing in an effort to reach my destination faster and thus expose myself to the risk of an accident. It is a well-known fact that if you do nothing you reduce the likelihood of making mistakes, but you also do not achieve anything positive. A prudent person knows how to handle freedom of choice, assess the risks involved and identify tactics that provide for safe actions.

7.3.3 Free Will

Deadlocks are a problem facing people, not the physical world. Clocks do not stop the way the human mind stops when it is unable to find ways out of difficult situations. Humans are doomed to try to find the resources that will enable them to achieve goals. In a sense, it is the price we have to pay for freedom of choice.

We have said that social organization is based on two types of interaction: *cooperation* and *conflict*. Cooperation is the coordinated execution of actions by more than one person in order to achieve a common goal. Conflict is the result of competition in order to obtain resources (see Sect. 5.2.2). Conflict may be internal or it may arise between individuals. Human societies differ greatly from insect societies, where cooperation is hard-coded into their genetics and conflict phenomena are rare. I have often wondered why evolution did not stop in societies organized solely based on automatic processes, such as those of insects.

I have explained that one theoretically interesting case is a conflict-free state. This is the case when there is an over-abundance of resources, and all possible actions can be realized independently from each other. Air, for example, is a critical resource for life. However, under normal conditions, no conflict arises since each of us can breathe air independently of each other. The same holds true for water and fuel although in this case their consumption entails a proportionate consumption of monetary resources.

Conflict-free systems are characterized by the *confluence* of their actions. In other words, if actions a and b are possible and do not conflict, then they are independent and I can carry them out in any order, reaching the same end result; this is not the case when there is conflict.

A deadlock with respect to a goal is a state where actions that allow for achieving the goal are not possible. Thus, bankruptcy may be a fatal development for achieving business goals, while the lack of freedom resulting from imprisonment limits the possibility of achieving social and operational goals.

Free will is understood as the possibility of choosing between actions by applying criteria resulting from the assessment of goals and the possibilities of achieving them.

There are two extreme schools of philosophical thought in discord:

- One is "deterministic," which asserts that our behavior is determined a priori by various causes and internal choice mechanisms that we cannot comprehend. This lack of understanding of causes is a source of an apparent uncertainty, which we resolve when we make decisions. In other words, for given initial conditions, the course followed by humans is predetermined. It is only human "stupidity" and lack of knowledge that make people believe they possess free will.
- On the other hand, existentialist philosophers raise strong objections against determinism. They (over)stress the ability of humans to make choices as one of our defining traits.

Determinism views humans from the outside as machines and refuses to confer an autonomous existence to mental phenomena, considering them a by-product of external actions, the outcome of physical laws driving phenomena outside of and within humans. Existentialism, on the other hand, views humans from within, acting of their own initiative to a certain extent and affecting their environment.

In my opinion, what exactly free will is and whether it "really exists" is a subjective ontological issue and is not amenable to rational analysis. Precisely what our choices are, whether or not they are a self-delusion, is of no importance. What is important is not just that our mind makes choices, analyses, and potentially resolves conflicts between actions, but, more importantly, that we experience the results of our choices.

Believing in free will is crucial with regard to our attitude toward life. We cannot deny that free will is a phenomenon playing a crucial role in how we handle all our relationships. People bear responsibility because they can choose. Machines bear no responsibility. Their choices have been resolved by their programmers. Of course, this does not stop certain foolish or ill-intended people from advocating for the rights of robots and artificial intelligence!

How can the concept of human freedom be defined?

In theory, at least, when there are no needs, the issue of freedom is not relevant. Many believe that freedom means to rid oneself of all needs or, at the very least, to minimize one's needs. This is why they stress that you have to even rid yourself of the idea of need altogether.

Needs, of course, are inherent in humans. I believe that freedom from a need does not mean that you are not subject to that need, but that satisfying the need is possible within the value system in place. In this case, needs may vary in terms of their nature and importance (physiological, safety, social, etc.).

A state of absolute absence of freedom is incompatible with human existence itself. However, I may face a lack of freedom with respect to the ability to achieve certain goals, for example, living a comfortable life, studying at university, ensuring that I remain fed.

It is harder to judge when we can be considered to be free. One first answer is as follows: when we have the choice of actions and meet their preconditions, and therefore have the resources, to realize a hierarchically structured set of substantive goals. However, caution is necessary here. With this definition, I must limit myself only to optional goals. I must not include choices of actions with mandatory or prohibitive goals that would be understood more as coercion than freedom.

The cost involved in achieving a goal must be commensurable to its importance. However, even if I do not take the cost of achieving a goal into account that does not mean that the more choices I have, the more freedom I enjoy. This is because, as we have already discussed, the human mind is limited by cognitive complexity (see Sect. 6.1.3) and cannot handle choices beyond a certain threshold. How this threshold is defined depends on the abilities of each person; however, it is certain that managing non-hierarchically structured choices beyond a certain number—let us say five—proves ineffective.

Analysis of a large number of choices at a state results in loss of time and, therefore, lack of effectiveness, if we wish to determine optimal solutions. An approximate analysis might be preferable, being briefer and allowing for more productive time to be used to carry out actions.

As is the case for computers, making choices also involves a computational cost for humans. We must always strive to reach a compromise between the computational cost of our choices and the effective and timely execution of our actions. I have seen people, mainly well-to-do, who have a plethora of options at their disposal and, being unable to manage them, live a life of confusion and unhappiness.

In summary, freedom of choice entails a computational cost, which we must limit for reasons of effectiveness. There are two crucial factors at play here.

1. One is the existence of a well-structured personal value system, which in a given state, allows us to choose effectively among a large number of theoretical options. It is no accident that the majority of ethical laws are proscriptive. There are people who, in order to live "peacefully," have imposed a large number of prohibitions on themselves, from minor to major ones. For example, one might not drink, not smoke, not drive, be a vegetarian, sleep every night at 10 pm and have a very limited budget for food and clothes per day. In this case, the number of choices definitely decreases. However, if a person feels comfortable living this way, they may be happier than a person who enjoys more freedom but is unable to effectively manage it.
2. The other factor concerns knowledge and its application in decision-making. Of course, the knowledge learned at school, university or elsewhere, is important for our ability to manage goals and prepare strategies. What ultimately counts, though, is not knowledge in itself but the ability to manage acquired knowledge.

The above framework is generic and can serve as a basis for comparing different situations. Even if a quantitative assessment does not make sense, there are clear qualitative criteria that enable us to say that one situation is preferable over another. Furthermore, the framework clearly indicates the role of ethics and ethical rules in particular to simplify the freedom game, as well as the importance of knowledge and "knowledge of knowledge."

In conclusion, I would like to discuss the balance between individual and collective freedom, which certain people often compare. I have noted that the social and political environment limits the degrees of freedom a person enjoys.

There are two well-known extreme views: individualism and collectivism. The former favors the unhampered action of the individual. The latter advocates for the subordination of individual choices and actions to common goals. At this point, I will not discuss what is obvious, i.e., that it is necessary to strive for balance between individual and collective freedom.

Imposing numerous proscriptive/mandatory rules in a social organization creates control mechanisms that hinder the performance and swift conclusion of social processes. This imposition also limits the degrees of freedom for the action of individuals, where they can exercise their initiative and inventiveness. Free spirits find it difficult to express themselves and be creative in a totalitarian society. I have witnessed how research productivity is limited in academic systems that are sclerotic and "hyper-hierarchical."

Mandatory/proscriptive rules must be minimal—provided, of course, citizens know that they must make optimal use of their freedom. We must allow for optional rules while also providing appropriate incentives for complying with them.

In the decision-making process, the present is a dividing line between the past and the future, what has happened and what will happen. The present is the privileged point where our conscious decision to act in a certain way or even to take no action shapes the future. It is the point where what comes to pass takes shape, where the past begets the future as the resultant of all actions, both conscious and subconscious. It is the moment when information is channeled into the system and free will is exercised.

Freedom is the conscious risking of the future and the existence of choices that will shape it in order to achieve certain goals. The price of freedom is the complexity of managing risks and choosing between conflicting actions. An absence of choices means that the future has already been predetermined by the past.

7.4 The Conscious and the Subconscious Self

7.4.1 General Considerations on Consciousness

7.4.1.1 Cooperation Between the Conscious and the Subconscious

I have discussed two systems of thinking, conscious and subconscious, and explained that they work together in a hierarchical relationship: conscious thinking is what normally overviews situations and issues orders (see Sect. 6.1.1).

Our perception of both systems of thinking is reflected, to a certain extent, in the distinction we make between conscious and subconscious. The conscious looks inwards; it "knows that it knows," it "sees" possible choices of actions in the mental system and makes decisions. The subconscious acts as an "auxiliary processor" to fulfil goals achieved through automatic functions. At this point, I must underline that I prefer the term "subconscious" to the term "unconscious" because, at least conceptually, it better describes the way I understand the role of the "not conscious."

I believe that our psychism was built gradually, like a game between our conscious and subconscious. The question is how the conscious builds the semantic model of the world that reflects the totality of our knowledge, and how conscious and non-conscious functions work together in order to make decisions and realize them.

The mind consciously feeds and teaches automatic thinking to perform functions that are essential for people, such as movement and speech. In fact, there are times when the conscious cedes control to the subconscious and momentarily deactivates itself in order to allow its unhindered autonomous functioning. When for example, we play football or go dancing, we loosen our impulses and become "lost," living an experience that is intuitive and transcendental.

It has been experimentally proven that we need at least half a second to consciously perceive and react to a stimulus. In contrast, the subconscious has a much shorter reaction time. In certain experiments, where a weak stimulus is followed by a strong stimulus over a period of less than half a second, the weak stimulus is grasped by the subconscious, while it goes unnoticed by conscious thinking.

It is remarkable how certain people exploit the two-way relationship between the conscious and the subconscious in order to influence our conscious judgment. I once read that a psychologist tested his students by giving each of them an analysis of their character based on their handwriting. Most of the students' judgments were positive, each of them saying how correct the analysis was. However, all students had received exactly the same analysis—which proves the text contained the judgments that each student would have liked to see, i.e., that they possessed robust intellects, with a few weaknesses and feelings of insecurity, but that they knew how to appear happy and strong.

Often, when faced with dangerous situations, the subconscious can take control so as to allow us to react automatically and instinctively, temporarily blocking rational, conscious analysis.

How do we strike a balance and self-control between the conscious and the subconscious? For an athlete or artist, the subconscious is a key factor for their performance. All transcendental experiences are characterized by this temporary loss of the conscious, whether they are religious, erotic, artistic, or relating to creativity, where the subconscious play an important role. An intervention by the conscious could lead to confusion and hesitation and damage the quality of performance.

You might go to a concert and know the music by heart. You could listen to it by playing a record at home. However, the result is not at all the same, as the physical interpretation is transformed into a host of visible and auditory information that carries you away. No emotion seems authentic when it is fully controlled by the conscious.

We must reassess the role of consciousness in our lives. Many of the wonderful things that people do go beyond conscious control. We can develop skills through intuition and practice in order to communicate with the world and access cosmic mystery. We can strike a fine balance between automated proscriptions and conscious choices. Automatic proscriptions help us overcome complexity, allowing the conscious to focus on meaningful and creative options.

7.4.1.2 Consciousness and Computers

We know that conscious thinking has limited information-processing capacity and accounts for only a small part of our behaviors and reactions. It is serial—we can focus our attention on just one object at any time. I have also explained that the complexity of relationships that the human mind can consciously grasp is limited.

Computers can obviously help us expand our intellectual capabilities. By their very nature, they can analyze large amounts of data, playing a role similar to that of the subconscious—that is, the role of "co-processor." We can thus enrich our knowledge and, indirectly, our mental capabilities.

The question is how human consciousness, given its inherent limitations, will be able to understand and control the development of an increasingly complex and rapidly changing environment (physical, technological). Unfortunately, humans are currently unable to grasp the dynamics of scientific and technological evolution and its impact. Systems have become so integrated and complex that it is no longer possible to monitor their operations analytically and to assess their role and usefulness in detail.

Consciousness is the culmination of human evolution. Therefore, while it created civilization that changed the world to a large extent, consciousness must not fall victim to its own achievement. Consciousness that does not realize its limits, i.e., that can only partially know the world, is a danger to itself. The civilization it created affects its own evolution. We need to understand this fact and help the collective conscience grasp the depth of changes and their consequences.

If you were to give technically sophisticated destructive weapons to primitive people who have learned to resolve their differences using bows and spears, you could cause a genocide—their collective conscience has not matured enough to measure the consequences of using such weapons in a heated conflict. During the twentieth century, it seems that we avoided a global nuclear war by gradually realizing that there would be no winners or losers in such a conflict. Averting an environmental catastrophe and becoming more aware of the rational and controlled use of computers are similar problems for our collective conscience.

We have confidence in big data and algorithms because they are supposedly better than we are at making predictions. When faced with the question "What should I study at university," we may consider the answers of a computer more seriously than our feelings and logic. When we are seeking a holiday destination or choosing among candidates to appoint as members of a committee, we may consider it more convenient to find answers using our computer. By making fewer and fewer decisions in our lives, we are fostering a spirit of indolence and lack of self-confidence. We feel relief by gradually shifting the weight of decisions onto computers.

As it would be unthinkable to build without tools, it will be equally impossible to manage the world without computers. However, the humanitarian vision that considers humans a central value will de facto be tested.

What will the golden rule be? Not to allow computers to make critical decisions by default. Not to allow, for the sake of "efficiency," the creation of closed and uncontrollable management and control circuits, of which only a few parameters can be regulated by us. For technology to remain in the service of humans, we should strive to rationally manage its complexity with increased awareness of our inherent limitations.

7.4.2 *Learning and Creation*

Humans learn through automatic mechanisms and, at present, surpass computers in this respect. The goal of education must be to foster not only critical thinking, but also skills acquired empirically. Football players, pianists, researchers learn through exercise and practice. Of course, during the learning process, they consciously apply rational rules, but inventiveness in problem-solving depends on knowledge that is difficult to analyze and theorize.

If a person can consciously solve a problem, they use acquired knowledge and, ultimately, a comprehensible method that can be explained, transmitted to, and used by others. If they solve the same problem without using reason and conscious thinking, I have to assume that an equivalent process takes place in their mind. In other words, there is an equivalence relationship, which at least holds true in logic and the theory of computation.

At this point, we have to take account of Bergson's views, which hold that direct experience and intuition are more essential for understanding the world than science and rationalism.

Indeed, while we can consciously analyze very little information (bandwidth), the capability of the subconscious to absorb and consolidate information is much greater. An image from a film, a painting, a line of poetry convey huge amounts of information through possible interpretations—so much that we will never be able to analyze them and understand what distinguishes a painting from a photograph. This is where intuition comes in, which is the possibility of grasping and synthesizing the "big picture" without resorting to analytical conscious thinking.

You can learn without profound theories, in practice, as they say. In order to learn how to swim or ride a bicycle, you are not given a method to study and then attempt. No matter how long someone explains it to you, you will not be able to swim in the sea based on the explanation alone because you need to learn how to automatically make certain moves without any analysis. When you learn how to play the piano, certain logical rules are "internalized" and become automatic.

You cannot learn through theory how to become a good researcher—just as you cannot learn how to become a good citizen. Of course, you can understand and consciously apply certain rules; however, this does not suffice. There must be assimilation into behavioral patterns. As the saying goes, "*la culture, c'est ce qui reste dans l'esprit quand on a tout oublié*"—culture is what is left when you have forgotten all you have learned—in other words, the result of thinking that has ultimately become automatic, subconscious.

This raises the subject of controlled liberation of the subconscious, allowing it to create masterpieces by subordinating it to a purpose. A great orchestra conductor, a creator, impelled by passion and willpower, can reach soaring heights. That is where the subconscious acts under the control of the conscious.

When you compose poetry, the creative process is largely subconscious, but the criteria for what is beautiful and grand have been learned by reading poetry and delving into the esthetics of language and meaning.

Artistic performance and creativity are intellectual achievements that require perseverance and effort.

A few years ago, I read in a book how Avyssinos, a renowned Cretan fiddler, whom I had the fortune to meet in person, learned to play the violin as a child. At first, he travelled from his village to the big city for a week to study with a music teacher. In the end, the teacher told him that, having shown him the basics, he had but to do as follows: "You will lock yourself at home for forty days, you will not go out at all and you will not do any work. You will just eat and sleep a little. You will play the violin day and night, constantly bringing to mind the melodies you know. As you play, try to carry them over to the violin. Once forty days have past, you will be a fiddler."

And it happened just as the teacher said. The young boy cloistered himself at home for 40 days and when he emerged, he could properly play numerous improvizations and local dance rhythms.

We may never come to understand exactly what the relationship between the conscious and the subconscious is through analytical processes. It is just like our relationship with computers. We interact, but cannot comprehend exactly how they function—that is, how information is represented and transformed.

7.4.3 Creation and Creators

I will conclude with a few personal thoughts on the process of knowledge creation. How are new ideas born? How do we arrive at solutions to difficult problems?

There is a special moment during the creative process where the mind becomes fertile and clear. Inspiration arrives like a spark, amazingly bringing solutions that we might have been seeking through hours-long logical analysis. The emergence of a solution possesses something magical. It is as if all our thoughts that might have gathered in mind in disorderly fashion, suddenly organize themselves and show us the way.

I recall what the great physicist James Clerk Maxwell said: "What is done by what is called myself is, I feel, done by something greater than myself in me."

It often happens that I wake up early in the morning with clear and creative ideas. Of course, I do not believe that they dropped from the heaven and came to me "free of charge." They are the crystallization of thoughts that had been tormenting me for a long while and now emerge in brilliant clarity. At times, I find it frightening; I wonder how I can do it. Of course, the satisfaction of creation is ineffable and wonderful.

It is also remarkable that, when I am considering a problem, my thoughts take root deep inside of me and I cannot get rid of them—to tell the truth, I do not even want to get rid of them—before I work on them in earnest and draw all the useful conclusions. It is like a disease, a virus that settles in the mind and will complete its "cycle" when I have fully explored the issue.

Creators are transformers of accumulated potential of experiences and knowledge. They listen and interpret, decode, and translate the ineffable. They have built their own world of symbols and representations and have their own paths that take them far away to the sources of inspiration.

I cannot but recall Socrates's "daimonion," the internal "voice" that gave him certain signs regarding his future behavior. A much more recent and characteristic case is that of the autodidact Indian mathematical Srinivasa Ramanujan (1887–1920), who died at the age of just 32 after making major discoveries in mathematics. A deeply religious Hindu, he believed that he developed his results based on ideas revealed to him by his family's deity.

I admire the insightful intuition—for lack of any other term—of ancient Greek philosophers. How did they make factual breakthroughs without experiments, without even having visual or other technical means?

What amazes me even more is that the great truths are not the result of thoroughly analytical or logically correct reasoning. The starting point is a driving idea (the importance of numbers, the imperishability of ideas, the synthesis of opposites, atomic theory, the continuous vs. discrete paradox, etc.) in support of which they seek any manner of arguments. I remember, when I was a high-school student, finding certain of Plato's dialogs to be dull. They presented interesting ideas, but at times, I had the impression that the thinking was incongruous or using flawed arguments. What I mean to say is that Plato's truly brilliant ideas stem from "intuition" rather than being the outcome of a purely rational process.

A typical example is Plato's conviction that planets, whose apparent trajectory in the celestial sphere is complicated due to the movement of earth, follow a normal and circular trajectory. The "argument," if I remember correctly, was rather esthetic—he could not fathom heavenly bodies behaving so irregularly! Another is the example of the young uneducated slave, from whom Socrates elicits the solution to a simple geometric problem in order to prove that the solution was already somewhere in the boy's mind, i.e., that learning is reminiscence.

How did ancient Greeks first lay claim to the freedom to think, to acquire knowledge? How did the idea of theory emerge as a logical reduction of the complex into a few predominant principles and elements?

In less than two centuries, we see a succession of intellectual giants, each of whom would add to the foundations of the temple of knowledge. However much one might disagree with the individual aspects, their conceptions are magnificent. Anyone who over the centuries—and even in the present day—has tried to approach major philosophical and scientific problems cannot but bring to mind these founding fathers of philosophical and scientific thinking.

In conclusion, I would stress that inspired creation is a lonely endeavor. Although creators may rely on the knowledge and help of a team, it is they who carve out the path, who determine the framework, the vision. I am not referring just to artistic creations, but also to designing an artifact, a major project which, even if it mobilizes thousands of experts, always has one "chief architect," a unit that inspires and motivates all the actors.

The ability to create is, in my view, something innate that can certainly be nurtured through education; however, knowledge alone is not enough. It is conquered through a process of deepening relationships between specialized items of knowledge and linking them to a more general consideration of the problems of a field of knowledge. It stimulates not only the conscious function of thinking, but also the creative and subconscious.

Access to deep knowledge is a torturous, solitary endeavor. Irony and arrogance are a veil before the eyes, putting us in danger of losing the way, leading to impasses. However, like an experienced explorer on a dangerous mission, creators must be vigilant so that they do not lose themselves in dead ends. I have seen talented researchers trapped in dogmatism, prejudices, over-optimism, underestimating the difficulties of penetrating knowledge.

Knowledge is conquered gradually. When you reach one peak, you see another, much higher peak in the distance. This is a game without end. If you ever feel that you have reached the end, you are done. You have lost the joy of ongoing creativity.

7.5 Seeking Happiness

Much has been written about what happiness is and how to pursue it. The aim of the discussion below is to say a few pedestrian and technical words relying on the mental model presented above.

I understand happiness as an individual's continuous capability to achieve his or her goals by taking reasonable risks. This is a dynamic game of the mind with the environment by developing appropriate strategies. It is dynamic in the sense that, when certain goals are met, other new goals may arise. Therefore, the pace at which goals are achieved must be commensurable to the rate at which new goals appear, so that the player is not overwhelmed.

We can also imagine that certain goals are shifting. The closer we get to them and partially achieve them, the more they become refined and enriched. For example, building a house or achieving professional success are goals that become clearer as we get near them and achieve them in part.

Happiness is to be moving toward reaching our goals by always remaining at some optimal distance—neither too far nor too close. The optimum is determined by the character of each individual. If I recall correctly, Kazantzakis said that each man's happiness is tailor-made to his measurements. There are those who do better with far-off goals, while others are better suited to closer goals. However, the goals must not be definitively reached because the game will then lose its appeal. I have seen people who felt professionally fulfilled—and they grew unhappy or even ill when they retired and suddenly withdrew from active living.

When you are too far away from where you would like to be, this becomes a source of discouragement and frustration. Of course, certain people cannot set long-term goals in their lives and are at the mercy of their immediate needs. This can happen either because the resources available barely suffice for their day-to-day survival or because they are incapable of mobilizing their intellectual and moral powers consistently and continuously due to mental indolence.

Happiness is directly linked to the ability to understand the world and yourself, but above all to create and envision. This is where education and background play a crucial role. If the mind is feeble and remains untouched by the joy of all manner of creation, then the game of happiness was lost before it ever began.

Although it may be adversely affected by subsistence needs and hardship, an individual's happiness is primarily a personal matter.

Collective happiness is not just the happiness of the individuals making up the whole. It can be defined accordingly as success in achieving common goals that are broadly and consciously accepted by the whole. Of course, collective happiness affects individual happiness, and vice versa.

In conclusion, I would stress that I am conferring purely technical content on the pursuit of happiness, and it has nothing to do with the moral values I will discuss in the following chapter. Kant says that morality is not about becoming happy, but rather about becoming worthy of happiness.

Happiness is a question of chemistry rather than logic. It is a balancing game between what you are and what you want to be. We are not satisfied when we feel inferior to what we would like to be. We say, "I don't want to be who I am – I feel ashamed of myself."

The same holds true in the opposite case, when we find ourselves at a loftier position than where we can or dare to stand, when we have to play a role that requires skills we do not possess. We sometimes say that someone falls short of a position they hold and that this is reflected in their behavior, which is insecure and stressful.

Finally, there is also the tragic case of a lack of self-awareness, where our self and our ego stand in ignorance of each other, i.e., we are not aware of our capabilities and weaknesses.

I once wrote the following:

"Know thyself." It is a prerequisite for harmony with ourselves in order to achieve a happy life. It is the first and last major imperative for humans. All the others follow, the way theorems follow from axioms. In our personal lives, knowing is not enough. Unmanageable knowledge can become an unnecessary burden that not only does not help, but also sometimes clouds our judgment.

Gaining knowledge through the game of freedom is the meaning of life. When the cycle of life closes, our conscience must have discovered and fully understood the rules.

Chapter 8
Value Systems and Society

8.1 About Human Societies

Human intelligence developed through various stages in which social organization played a key role.

The words *community* and *communication*, both of Latin origin, are clearly related. There can be no society without communication. The intelligence of living beings evolved, thanks to social organization.

Societies are complex information systems.

Insect societies present impressive collaborative behaviors. They are characterized by a limited number of character types. For example, in bee society, these comprise workers, drones, and queens. Each type has well-defined patterns of behavior and communication, which determine their ways of interaction. Thus, what characterizes insect societies is their collective behavior, with emphasis placed on cooperation to achieve common goals of collecting food and reproducing. Insects probably have no awareness of dilemmas and conflicts. They do not exhibit "selfish" behavior.

In the case of mammals, we are moving onto a different evolutionary stage, where animals have a strong feeling of self-preservation, which entails selfish behaviors and "dog-eat-dog" conflicts. Such phenomena are intense in primates, where violent conflicts are frequent and are limited only by fear of the stronger.

Human societies stand out from those of all other animals in that, through evolution, they have developed organizational structures that allow for the coordinated achievement of complex goals by defining the framework of actions of individuals in the structures in question for their achievement. I have already explained the importance of a common value scale and regulatory frameworks for social organization.

We can regard human societies as complex information systems that possess all the main characteristics of a mental system since they are theoretically the result of a synthesis of individual mental systems. Of course, this synthesis takes place within

J. Sifakis, *Understanding and Changing the World*,
https://doi.org/10.1007/978-981-19-1932-9_8

structures and is regulated by the common value system. Thus, the analysis in Sects. 7.2 and 7.3 can be applied to a large extent to social groups, the difference being that the individual's consciousness is replaced by a more abstract concept: *collective consciousness*.

Studying the dynamics of human societies is extremely complex. This complexity is not just numerical, as is the case in physics when studying a large number of atoms with exactly the same characteristics. It is "disorganized complexity" (see Sect. 4. 1.3), which depends on two factors.

On the one hand, each individual in society as a whole has particularities, which cannot be summarized in a "type of individual," as is the case, for example, with the atoms of a gas.

On the other hand, individuals hold different positions in the social organization, for example, due to income, occupation, or office, which affects their behavior. This complexity makes it particularly difficult to study social phenomena, which exceeds the aims of this book.

In this chapter, I will discuss how three important value scales take shape: gnoseological, ethical, and economic, using concepts from previous chapters. I will explain how the composition of each individual's subjective experience takes on an objective dimension, which can be studied as a social phenomenon. I will discuss the institutions that play a fundamental role in shaping and maintaining the scale of values. Finally, I will analyze the phenomenon of the deconstruction of values and the decline this leads to, ending with a discussion of democracy in light of the foregoing.

8.2 Gnoseological Values

Gnoseological values concern the validity of our knowledge about the world (epistemic values), on the one hand, and the possibility of applying knowledge (technical values), on the other. These are common social values reflected through the filter of subjectivity on individual value scales. Consequently, gnoseological values can be judged according to objective criteria, as well as in relation to their degree of acceptance by society as a whole, which I will discuss in what follows.

8.2.1 Epistemic Values—Truth and Falsehood

A key property of knowledge is that it is "true"—otherwise, it would lose its utilitarian value—i.e., it helps us understand and change the world (see definition of knowledge in Sect. 3.4.1).

Truth is an epistemic value and characterizes the concordance with reality and the rules of reasoning. *Falsehood* is therefore defined as the opposite of truth. Here we must make it clear that we usually confuse falsehoods, being "untruths," with lies, which are a conscious distortion of the truth. An untruth characterizes information that is not empirically or logically substantiated, while lying is an ethical value.

A person knowingly telling a falsehood is lying and thus has a moral responsibility. The phrase "the earth is flat" is false. If someone in the Middle Ages were to claim the phrase was true, then they were possibly not lying, because it reflected their level of knowledge. In contrast, if a present-day educated person were to assert that the phrase is true, then they would be lying.

Nevertheless, we cannot certainly ascribe a value of "true" or "false" to every sentence that constitutes information regarding the world. This is generally due to our inability to precisely define the meaning of natural languages. When we talk about social or economic phenomena or express judgments about people's behavior, our propositions may be open to multiple interpretations and it is therefore impossible to ascribe the value of "true" or "false" to them.

However, even in the field of mathematics, where language is strict, there are renowned undecidable propositions, which we discussed when referring to Gödel's theorems (see Sect. 3.4.2). There are also inherent limitations to knowing the truth about physical phenomena, as we have explained in detail.

The truth has both an objective and a social dimension. Its emergence takes place through a complex social process. What a given society considers to be "true" reflects the level of knowledge at the time, shaped through filters controlled by social institutions. Perhaps someone would say that I am kicking at open doors by stating the obvious. Everyone knows the story of Galileo and the other "heretics" who came into conflict with established views. In this case, the emergent truth challenged the authority of the Church.

Every kind of truth, in order to be adopted by the many and become a social value, must be subjected to a process of certification, where the social institutions play a crucial role.

Let me use a completely "neutral" example as an explanation. In the mid-1990s, there was intense debate around the proof of French mathematician Fermat's last theorem. This is an extremely difficult problem that remained unsolved for 358 years, despite many asserting from time to time that they had solved it. In 1995, the British mathematician Andrew Wiles published a correct proof. How is it that we now believe that this problem has been solved? This is because groups of mathematicians—whether or not organized, it matters little—have read the long and highly technical proof that Wiles published and agreed that it is correct.

Similar processes occur with all kinds of emergent knowledge. In order for the theory of relativity to become common knowledge and established, it took debates at academies, publications in journals, books, and newspapers. This is how it became knowledge as part of the educational system and entered the public domain.

Unfortunately, in the knowledge and information society, institutions such as the press and academic organizations do not play the role they should be playing as custodians of truth, nor do they help properly inform the public of all the major problems of concern. This poor and sometimes distorted provision of information poses a huge risk to individual and social liberties.

Increasingly, fabricated lies become a tool to mislead and manipulate the many. The powerful falsify or hide the truth, spread false rumors, use rhetorical figures to impress and mislead. The internet and social networking are multiplying their devastating effects.

This is particularly the case when people, being listless and disoriented, do not care to protect themselves. In fact, people have the natural tendency to believe anything unbelievable and absurd that stirs up emotions and passions, which cultivates illusions. They close their ears to the truth and real problems that disturb and upset them. At present, the concept of truth is at risk due to confusion and familiarization with falsehoods, due to the dulled judgment of the many.

Finally, I would like to say a few words about what has been called "post-truth." The concept describes the lack of common criteria as to what is true and what is not. It states that fiction counts more than facts. Why is this?

I think that what we are experiencing is the result of two factors in particular. On the one hand, it is the moral crisis and relativism, which blur the boundaries between truth and falsehoods. On the other hand, it is the fact that public opinion is being bombarded by a wealth of information of all kinds, and that it is difficult to discriminate between unsubstantiated "news" and corroborated facts.

Public opinion tends to uncritically believe rumors and fabricated news. People are fed up with and saturated by the "real stories" served up by all types of media on a daily basis. Promises that are not kept, manipulations and deceit lead to "irrational" behavior.

How can we explain Brexit or the Capitol attack in January 2021? How can we explain the boom in religious fundamentalism and religious conflicts that have been testing Western societies since the late twentieth century? How can we explain the powerful anti-vaccination movement being observed in some European countries, placing the lives and liberties of thousands of people at risk on unfounded and fear-based arguments?

Societies are simply tired of the parody of modern democracies and their inability to solve blatant problems. It is a failure of the elites to shape and realize a vision for stability and prosperity.

How long will this situation last, and how will it evolve? It is difficult to predict whether this is a temporary absence of rational thought or a permanent triumph of the absurd.

8.2.2 Technical Values

8.2.2.1 Correctness and Erroneousness

While the truth involves concordance with facts and logic, "acting correctly" means taking a step toward achieving a goal, which is defined by specific needs. An error, therefore, is any deviation from a goal. It is integrally linked to our mind's inability to properly assess a situation, to aim right.

"Correct" and "erroneous" are technical values. Error etymologically originates from the Latin verb "errare," which means to roam, i.e., to stray from the right path.

Technical values help us assess whether an action is successful according to gnoseological criteria. They also allow us to weigh the responsibility of the actor: this is the case when an error on their part causes damage and they are found to be negligent because they did not properly apply their technical knowledge. Thus, any expert, whether an engineer, a doctor, or a cook, may make mistakes when applying their knowledge. These result in professional liability, for example, when a building designed by an engineer collapses or a surgeon endangers the life of a patient due to a manifestly incorrect action or omission.

While what is true is assessed according to objective criteria, what is correct is judged in relation to goals and their feasibility. These goals must be clearly formulated and, of course, must not contradict the legal and ethical rules in force. In any case, knowledge of the truth is a prerequisite for us to act correctly.

The most assured way to make no errors is to do nothing. Both the prudent and the foolish make mistakes. However, prudent persons have the foresight to anticipate the impact of their errors and to take measures to prevent critical situations. They may also deliberately make minor mistakes in order to investigate a situation and learn—"paying to learn," as is the case in poker, in order to understand the opponent's tactics. However, prudent people make sure that their errors are not irreversible.

As important as the truth is, so is social highlighting and certification of what is correct. That is why there are institutions that play a controlling and validating role. We board airplanes, live in skyscrapers, use electrical appliances with the conviction that they are quite safe because they were designed based on theories and rules; because certain independent authorities have certified that they have been manufactured according to correct procedures and specifications.

I have pointed out that, while standardized certification procedures are applied for each product, such as children's toys or electrical tools, there are no such procedures for computer systems and their services. No certification is required to ensure that they function properly with respect to specific safety and security rules, except for certain critical applications, such as transport, nuclear power plants, and smart cards. This situation engenders the risks already discussed in Sect. 6.4.2.

8.2.2.2 Risk Management Principles and Their Implementation

I explained that managing goals is linked to the concept of risk, which characterizes the probability of finding ourselves in states that range from hazardous to catastrophic, with high value costs (see Sect. 7.3.1).

Let us use the coronavirus pandemic as an example to discuss risk management principles. The different way each country has handled the pandemic highlights the complexity of the problem of assessing risk and reaching compromises between safety and liberty (see Sect. 7.3.2).

There are two ways of approaching risk management.

1. An approach is based on the *precautionary principle*, which "requires a rapid response to be given in the face of a possible danger to human, animal or plant health, or to protect the environment," as described in European legal texts [1]. This approach is appropriate in critical situations. Where the scientific data do not allow for a full assessment of the risk, radical measures must be taken on the basis of the "worst-case" scenario. The principle is applied in cases of catastrophic weather phenomena, epidemics, and other natural disasters. It is also applied in order to assess the risk of using pharmaceuticals, medical devices, and means of transport, the development of which is subject to control by independent organizations. The precautionary principle is based on another, more fundamental principle governing the European legal framework: that life is of the utmost value. When a human life is at risk, the cost required to save it is unimportant.
2. The second approach to risk management is applied when the criticality of what is at stake is minor or, at the very least, does not concern human lives or severe environmental disasters. It concerns the *optimization of criteria*, where conflicting factors that characterize an activity arise. If, for example, the risk (i.e., the probability of something harmful occurring) increases in proportion to the intensity of an activity, then I can determine, if I have the appropriate mathematical model, the optimal level of activity. Therefore, if I am driving on a motorway, there is, theoretically speaking, an optimal speed to reach my destination which is a compromise between travel time and risk. The optimum depends on many objective factors such as weather conditions, traffic, and car type, as well as subjective factors, such as driving ability. The optimum corresponds to the level where the benefits are maximized in relation to the associated costs due to risk. Finding optimal operating points for business risk management is one of the fundamental problems of economics.

These two ways of addressing risks are reflected in the management of the coronavirus pandemic. At first, the leaders of many countries behaved more or less as managers, taking the second approach. Human life was simply the adaptation parameter of an equation, where stock market, economic and political criteria were taken into account.

The application of the precautionary principle would require the immediate imposition of strict measures aimed limiting the spread of the virus so as to save as many lives as possible. However, this was not the case except in very few countries. Thus, while the contamination of the population in the United Kingdom was out of control, the British Prime Minister was talking about "herd immunity." We understand that he prioritized keeping economic activity at a satisfactory level at any cost. He did not take into account the cost in human lives when 60% of the population would have been infected. Therefore, even if a small proportion of patients lose their lives that translates into tens of thousands of lives.

A similar attitude was shown by the USA, where President Trump announced that the threat to "the population and economy" was minimal and referred to a "Chinese" virus. It is interesting to note that fear of the pandemic's impact on the economy was at least as significant as fear of loss of human life. The then Governor of New York State, Andrew Cuomo, made statements against Trump on 28 March 2020, claiming he did not have the right to put the city in quarantine, characteristically stating that "New York is the financial sector. You geographically restrict a state, you would paralyze the financial sector."

However, even in other countries closer to us, governments were unacceptably flippant, adopting a "wait and see" attitude. France held the first round of municipal elections on Sunday, 15 April 2020, and was forced to immediately cancel the second round the following day.

Given the foregoing, I would also like to highlight a dangerous change in attitudes to crisis management, which was gradually adopted by the centers of power over the course of the last few decades. This marks a transition from strict adherence to precautionary principles in the 1970s to adaptive management, where human life is not what is most important, but just a parameter in a game dominated by economic and political criteria. Unfortunately, we moved from a time when at least the major Western democracies were quite robust and had the means to implement their policy, to an era when even the policies of the most powerful state are tied to the whims of the markets.

Similarly, with regard to system security issues, I have had the opportunity to see this slackening and transition from the precautionary principle to a financial-style risk management, even when human lives are at stake.

I have explained that the USA currently accepts self-certification for critical systems. This means that there is no safety limit certified by an independent and responsible authority, but that this limit is determined by the manufacturer itself. One characteristic case is that of the Boeing 737-Max, which had been self-certified with a single angle-of-attack sensor, while regulations require at least two sensors [2].

This policy change was considered necessary because the precautionary principle makes the widespread use of new technologies, such as artificial intelligence and autonomous systems, economically prohibitive.

The way the coronavirus crisis is being managed is indicative of the revival of a systemic ideology of barbarism at the global level. It highlights the weaknesses of a system, which are increasingly visible in crisis management, such as climate change, but also control of the use of new technologies. The exclusive priority of the economic and technological factor irreparably jeopardizes the foundations of our civilization, the cornerstone of which is respect for human beings and their lives. I am afraid that the price of barbarism will be even heavier in the future.

8.3 Ethical Values

Ethics is the area of philosophy that studies the concepts of *good* and *evil.*

Ethical values are crucial for social organization and are therefore assessed on the common value scale; at the same time, they are also individual values. They would be completely useless in societies without friction and without uncertainties. They contribute to conflict resolution and achieving consensus, mutual trust, and predictability, which are essential for social peace and progress.

Complying with ethical rules is a matter of choice and therefore depends on each individual's sense of responsibility. Sometimes this stems from the need for recognition by society as a whole when society values it and rewards such compliance.

Throughout history, modern societies have adopted a relatively comprehensive set of ethical rules and values.

Ethical values emerged alongside the first human societies. Initially, their development was inherent in the development of religious sentiment. Later, ethics became the subject of rational study by philosophers, starting with Aristotle.

Contrary to gnoseological values, ethical values cannot be based on objective criteria. Their adoption is the crystallization of best social practices or is a question of faith (belief)—but this does not mean that ethics lacks rational foundations.

In Sects. 8.3.1 and 8.3.2 below, I will discuss the concepts of good and evil and the role of faith in laying the foundations of ethical rules and managing ethical choices.

8.3.1 Good and Evil

Good and evil are defined as right and wrong, respectively, in terms of ethical values and rules. Managing good and evil is the central ethical problem facing our conscience. Any breach of ethical rules is considered to be bad, for example, non-compliance with mandatory rules or violation of proscriptive rules.

Usually, there is some distance separating good and evil. They are concepts that, while defined by subjective criteria, also have a broad objective basis in the common value system.

It is no accident that the golden rule of ethics is common across many peoples: Cleobulus of Lindos wrote "*ὃ σὺ μισεῖς ἑτέρῳ μὴ ποιήσῃς,*" while the Old Testament includes the phrase "*ὃ μισεῖς, μηδενὶ ποιήσεις*"—both essentially meaning "do not do unto others what you do not want them to do unto you.". This is an admirable empirical rule of balance that ensures harmonious coexistence in societies.

Many ethical rules have a deterrent or prohibitive character, such as the rules of the Mosaic Law, "thou shalt not murder," "thou shalt not commit adultery", and "thou shalt not bear false witness." These rules are also of practical interest: they reduce individual choices and complexity in managing freedom.

In theory, ethical rules seek to ensure that people co-exist harmoniously and work together in a given social context. As such, they are not independent of religious and other beliefs. Furthermore, they tend to shore up the status quo and the rights of the powerful.

In the Middle Ages, "buy cheap and sell high" was a grave sin, constituting moral misconduct and, at times, even a criminal offense. Seeking riches for the sake of riches was reprehensible, and flaunting one's wealth drew the ire and condemnation of society.

Today, seeking profit by all means is regarded as evidence of gumption and, to a certain extent, a virtue. The nouveaux riches of recent decades in particular, not only do not hide their wealth, but also appear proud of their power and their vast fortunes.

Certain ethical rules aim to establish relationships of trust that allow for the predictability of behavior. This is absolutely essential for the convergence of the actions of individuals in the pursuit of common goals and the effectiveness of transactions. Such rules include not lying, keeping one's word, not being hypocritical, etc.

Other rules require that our actions be inspired by a sense of fairness, i.e., not to put our personal interest above all else, ignoring the impact on others. This means courtesy, respecting the choices of others, not unnecessarily excluding them and disregarding them.

When we are selfish, even when pursuing a small personal interest, we can harm others by hindering the achievement of common goals. In contrast, when we are being altruistic, when we overcome our personal interests and, in certain cases, when we make personal sacrifices, we contribute to others and society at large.

Other ethical values concern how we choose actions, such as a sense of responsibility and courage. You are responsible when you choose actions that not only do not violate ethical rules, but also lead to the best result. There are many who do no harm, but few who, out of loyalty to and belief in the common good, try to do what is best at every turn.

You are being courageous when you have confidence in your strength, self-discipline, optimism, and perseverance in achieving goals.

Evil violates ethical principles, while anything that does not run contrary to fundamental ethical principles and contributes toward individual and collective happiness is considered to be good. Of course, both depend on the circumstances and have range and intensity. They are judged on results in the short and long term. A small good can result in a big evil over time and vice versa. One major difference is that good generally takes effort, while evil can also be done when one does nothing. Neglect and inertia can cost dearly, as time waits for no one.

A virtuous act is one that results in positive ethical values as defined by the common scale of values. Conversely, the absence of virtue characterizes actions that result in negative values. Thus, we speak of cowards, egoists, criminals, etc. As I said, selfishness is manifested through actions that, above all else, lay down criteria for meeting personal needs, ignoring the effects of these actions on others. Acting illegally may be ethically reprehensible but is characterized as non-compliance with the law.

The absence of virtue at the ethical level leads to a reduction in self-esteem, which can manifest itself in feelings of shame or guilt in certain individuals. Shame is the feeling that we did not rise to the occasion and violated ethical rules. Guilt concerns acts that are usually considered criminal by a legal system. Guilt may result in remorse, a feeling deriving from fear of the punishability of the actions in question.

Often the pursuit of good can bring us into conflict with others, especially those who pursue evil. This is the case with the heroes of antiquity, the prophets of the Old Testament, and the enlightened individuals we celebrate as martyrs. Hercules fought evil during his 12 labors, just as Theseus punished the brigands and freed the Athenian youths from subjection to the Minotaur. Martyrs sacrifice themselves, actively resisting evil by fighting for human and national ideals or by not accepting to yield and deny the ethical values dictated by their faith.

It is our moral obligation as human beings, not only to do good, but also not to give in to evil.

I must point out here that there are certain gray areas. A "good evil" idea is firstly an evil one, i.e., it has harmful consequences or engenders risks, but in some circumstances, it becomes so attractive that its glamor eliminates the inhibitions regarding its execution. Examples include stealing from the rich to give the poor, breaking things during a demonstration in order to claim what is right. I believe that ethical goals must only be achieved by ethical means. The opposite approach can lead to dead ends and vicious circles.

"Evil good" ideas are more dangerous. They are popular, enchanting ideas whose glamor can seduce us into hurriedly and uncritically realizing them, without thinking about what their realization might cost or whether they are even feasible.

These ideas set goals that, in theory, meet high ethical criteria, such as absolute social justice, and whose implementation leads to disasters or dead ends. These are "blanket" ideas, which in the name of misconceived equality, limit creativity, and individual initiative and do not accept the diversity of individuals. They feed populism and demagogy.

We can find numerous examples of "evil good" ideas in all aspects of life, especially in political and social life, such as educational reforms everyone is in favor of, debt forgiveness, and early retirement at the age of 50. However, this is also the case even in scientific research, where vast amounts are spent uncritically on ostensibly lofty but manifestly infeasible goals, or where the expected benefit is not commensurate with the cost entailed.

The relationship between the conscious and subconscious in the sphere of ethics is also interesting. Once a good Israeli friend, who unfortunately left this plane of existence early, made the following interesting statement: the actions of a good Jew need be in line with the Laws and Commandments; how he thinks is indifferent in God's eyes. There is a difference here with the Gospel message, which condemns both the sinful act and the thought: "And if thy right eye offend thee, pluck it out, and cast it from thee," "And if thy right hand offend thee, cut it off, and cast it from thee."

This difference had long since given me pause, and I believe I have an answer regarding its importance. For a Christian, doing good must be a natural disposition and not an obligation, it must be something automatic and spontaneous. This is because their conscience has nourished their soul and the Christian does good as easily as breathing, as naturally as their heart continues beating. You do good not just to show God that you are following His law. Doing good becomes a need, not a coercion. It is inherent to your existence.

US President Jimmy Carter once fell into a trap, during an interview on the campaign trail, when asked whether he had ever cheated on his wife. He answered:

"I've looked on a lot of women with lust. I've committed adultery in my heart many times." This stirred up a storm of disapproval and protest!

There are strong desires that have been pushed down to the subconscious and no longer subject to willpower, for example, when you decide to quit smoking permanently or decide to become a vegetarian. These automatic rejection mechanisms, which we create at a time through our willpower, simplify the game of choices. This can make our life easier or more pedestrian, depending on the situation.

8.3.2 Faith and Dogmatism

"How can a scientist be a believer?", someone once asked me. He happened to be a mathematician, so I reversed the question, asking him if he understands why axioms are needed in mathematics. I said that my beliefs concern questions that are beyond the realm of knowledge. Believing does not mean that I do not accept the scientific approach. Belief complements knowledge and is necessary for achieving coherent behavior. He insisted and was absolutely convinced that any kind of faith is a prejudice from which all rigorous thought is naturally free.

Faith is a concept that has been completely degraded in these times. This may well be because it became worn out over the centuries, due to our having loaded it with much more than it could bear. Then again, perhaps it is because we are living in a time when opportunism is considered a virtue.

A twentieth-century French politician, who stood out because he occasionally served various political ideologies, said in jest "it is not the weathercock which turns; it is the wind!"

If we consider the a priori acceptance of certain beliefs as faith, then no approach to knowledge is free of this consideration. Mathematics is based on axioms. The development of any scientific theory is based on hypotheses that, of course, must not be contradicted by reality.

Faith in ethical and spiritual values can be considered a set of beliefs, a type of "axioms" that cannot be directly validated. However, we are able to judge them by the result of their implementation. Even religious beliefs and the resulting practices are tested through personal experiences. Each individual possesses a metaphysics, whether they admit it or not.

As I have explained, the mind uses the individual scale of values and relevant rules as knowledge for assessing needs and making decisions. You cannot build a logical construction without points of support. Without such support, the mind flounders and gets nowhere.

Therefore, adherence to ethical principles also stems from a practical need to effectively manage one's liberty. We can confer a purely technical character upon this adherence.

What we call dogmatism is something different altogether and has nothing to do with adherence to values. Dogmatism means giving your convictions a basis, which is unquestionable and cannot be debated, and trying to impose these convictions at times. We must fully respect the convictions of others, insofar as they do not establish any infringing practices or offend public sentiment. Therefore, freedom of religion must be a right in every society founded on rule of law.

In conclusion, I would say that it is better to have a wrong faith than a lack of points of reference and an ambivalence. An intelligent, honest person will discover in practice that adherence to the wrong convictions creates practical problems and will revise them accordingly. We must not look at things in a static manner. Adherence to ethical and spiritual values and faith in the common good is not the result of divine inspiration. It is not something that someone places in your head and stays there once and for all. Convictions are created or lost, are jeopardized and challenged every day in the dynamic collective game involving people's actions, behaviors, and relationships—political, economic, social, from the simplest to the most formal.

8.4 Economic Values

Distinct from that of all other species, human social organization is based on the cooperation-conflict model. Economics considers human behavior as a relationship between given needs and limited resources with alternative uses. Needs and resources are matched through market mechanisms, which dynamically determine economic values by means of the interaction of supply and demand affected by consumer behavior trends and the role of institutions.

The study of the economy, as we know it, dates back to the origins of capitalism. Before economic laws could be understood and implemented, many changes had to take place.

At first, economic relations were defined by traditions and religious prejudices, which aimed at maintaining harmonious coexistence in societies. For example, the caste system in India aimed to strike a balance between professions. A similar stratification system was employed in ancient Egypt and in ancient China. The idea of profit for profit's sake was deplorable in the Middle Ages. For centuries, guilds played a major role in urban life and exercised a stranglehold on market economies by protecting their corporate interests.

Money is one of the most powerful abstractions of the human intellect. Every unit you acquire is a promise that you will have goods and services, that others will offer so that you can purchase anything for sale in the future. You can therefore purchase confidence for your future consumption and freedom of time, but you cannot purchase a future with your money, that is, the actual time to live your life. This is—fortunately—a fundamental limitation of money. It would be a nightmarish world if one could buy life, for example, take another's life counter down to zero and add it to one's own.

What I want to stress is that the economic game has an important informational dimension, where ethical and technical values are crucial. The effectiveness of social organization is equally crucial. It depends on mechanisms for establishing ethical values, the legal and regulatory framework for transactions, the processes of wealth creation and redistribution, and the management and development of knowledge.

Transactions require mutual trust to reduce risk between the parties involved and to enable seamless and secure communication, which is currently achieved through authorized bodies. This view is confirmed through the rapid growth in transactions carried out via computers and telecommunications. It negates the one-dimensional standpoint that only emphasizes the strictly material nature of exchanges.

The first banking systems were developed in the West during the Crusades and subsequently through the Jewish diaspora. The existence of institutions to facilitate and guarantee transactions between third parties and thus profit is crucial for the development of commerce and finance. Central banks, through their dual mandate as independent authorities, shape monetary policy and oversee banking institutions.

In the future, the use of blockchain technologies will allow for transactions between parties without third-party guarantors. This may threaten the existence of banking institutions causing them to surrender centralized control unless they adopt blockchain technology, particularly in order to address complex cost and operational challenges.

However, the lack of central arbitration and guarantee bodies entails high computational costs. It also makes transactions more vulnerable to cyber-attacks. That is why it is difficult to predict what the future of these technologies will be. If they ultimately replace central banking and control institutions, this will not just seriously impact the organization of transactions at the international level but will also be an important step toward deeper globalization.

I will end with a few thoughts on the evolution of the dominant economic model toward what we have been calling neo-liberalism since the mid-1980s. This call for "less state" in the productive sector brought about a series of reforms in the Western world accompanied by massive privatization, with the state being reduced to playing a purely regulatory role so as to allow the "natural," smooth functioning of the market, at least during growth periods.

Most states no longer have sufficient financial resources of their own and drivers necessary to intervene in a forward-thinking manner. They do not have a "nest egg" for difficult times and must raise funds by borrowing from international markets. In my view, this premeditated weakening of states will serve as yet another step toward broadened globalization and the disappearance of the "nation state." Countries such as China and Russia, which are protected by a kind of state capitalism, may escape this trend.

At the same time, we are experiencing the internationalization and integration of commercial and financial relations, thanks to the opening up of markets and the computer revolution. This led to the rapid growth of big tech companies such as Google, Apple, Facebook, Amazon, and Microsoft, which boast enormous economic power and influence. The global dimension of these businesses has allowed them to escape the control of states in terms of paying taxes and complying with regulations and laws.

Neo-liberalism goes a step beyond slackening social ties and the rejection of any idea of interventionism and programming. It strives for competition, the creation of inequalities, and the reduction in social services. All this as a means for creating markets, which entails destroying collective structures and fostering a culture of individualism.

This leads to a deification of the market and its automated processes.

Humanity has suffered from the economistic view that all social phenomena can be analyzed based on the economy and its dynamics, focusing on the economic game and disregarding the contrasts between the individual and collectivity, humans and nature, nations and races. Unfortunately, this spurious view serves as a meeting point for both historical materialism and neo-liberalism.

Neo-liberalism proposes a kind of "cybernetic" vision of a market-driven world: efficiency is achieved through the automatic interaction of economic "machines" that exchange information and are motivated by profit. Von Hayek, one of the instigators of neo-liberalism, says that the market is an information processor more powerful than the human brain. I quote: "the market is posited to be an information processor more powerful than any human brain, but essentially patterned upon brain/computation metaphors" [3].

I cannot know where such worship of the market can lead humanity. However, I am not optimistic about the outcome.

8.5 The Role of Institutions

Since their inception, societies have been based on the existence of institutions. These are structures that contribute to the imposition of social order by promoting the common value system and the application of explicit or implicit rules. Explicit rules are laws, regulations, and institutional frameworks that prescribe the actions of individuals and bodies. Implicit rules characterize the attitudes and morals of the social environment. They play a major role in the successful implementation of the explicit rules. One characteristic example is the attempt to apply the US constitution in Latin American countries, which failed miserably due to the different social environment.

Through their actions, institutions define, to a large extent, the scale of values of a society: what is true and what is not credible, what is right and what is wrong, what is legal and what is illegal, what is good and what is bad. They codify balances, which take account of the social and political conditions that have developed over the course of history. Their proper functioning and adaptation to shifting situations are essential for the survival of a social system.

For each field of action—economic, political, legal, educational, military, gnoseological, ethical, religious, esthetic, etc.—there are corresponding institutions that define and promote specific purposes and values, which all work together to achieve social order and some idea of the common good.

In primitive and ancient societies, interactions in relatively small social groups were personal. Misconduct brought shame to the perpetrator and their family. Such relations, which were still essential in democracies such as ancient Athens, were definitively lost upon the disappearance of the city-state and the creation of empires, such as those of Alexander the Great or the Roman Empire.

In modern societies, the state has a responsibility to renew and adapt existing institutions, and to develop a vision that will safeguard the future of society as a whole by planning actions for growth and prosperity.

Institutions form public opinion to a large extent and foster social cohesion and synergy. What is the state of the world economy? How does a country's education system perform? Is artificial intelligence dangerous? How are we dealing with climate change? Is a vaccine dangerous and, if so, how much?

History is full of examples where misinformed or insufficient public awareness is swayed into making wrong decisions with devastating results.

We see that the current crisis of values reflects and is reflected in the inability of the institutions of national and international organizations to rise to the circumstances.

One blatant example is the failure of the United Nations and, more recently, the World Health Organization, to respond effectively to crises with a view toward the common good. At the same time, the rapid development of science and technology has made some believe that we will live in an increasingly better and more controllable world. However, progress alone cannot guarantee that human societies will be able to adapt and effectively address the challenges emerging, which are increasing automation, overpopulation, pandemics, and climate change. International cooperation and the existence of effective international institutions are more necessary than ever.

8.6 Depreciation of Values and Decline

How does a society "decide" to die? How are nations, civilizations, and empires lost? How does a society break down into, but a patchwork of people impelled by their tendencies and appetites with no internal cohesion, easy prey for malicious and predatory people?

There are numerous examples of societies that declined and collapsed not so much due to external risks or economic crises, but because they ceased to function as coherent social groupings. One characteristic example is the decline of the Roman Empire, which was due to a crisis in ethical values, and the progressive degradation of the value system on which the *res publica* was based.

Societies function well as long as there is strong cohesion around a consistent, common scale of values, as long as individuals consent to sacrifice part of their personal liberty in order to achieve common goals.

Every social crisis is the result of the depreciation of the common value system in its various fields. It is not easy to understand how fields affect each other. Depreciation is the result of a general slackening, a mismatch between established values and newly emerging ones. Of course,—a rare occurrence—the crisis can prove fruitful when the emerging values form the basis of a powerful vision for change, such as the French Revolution.

It is interesting to examine how certain countries that had all the qualifications for growth and prosperity, declined and suffered a profound economic and political crisis. One characteristic example is the case of Argentina, a country vastly rich in natural resources that experienced an exceptional period of prosperity and growth comparable to that of advanced countries in the first half of the twentieth century. Since then, it has fallen into a cycle of successive crises and decline.

More or less similar situations show that factors relating to social and political organization, the functioning of institutions and the cultural level of people are more important than material goods. Many underdeveloped countries have the material conditions for progress and prosperity. Economic determinism, which places predominant emphasis on the economic factor, has failed miserably.

Even more important than economic resources is the ability of a society to act as a coherent (information) system capable of effectively bringing its creative powers together around realistic goals and planning to achieve them.

8.6.1 *Decay of Language—Degradation of Concepts*

Values start being undermined when language and concepts are degraded. If a language grows poor and concepts decay, we lose the corresponding possibilities of formulating knowledge.

Language is one of the most valuable common goods. Its erosion opens the path to decadence and decline. When words lose their common conceptual value, they become like a useless, counterfeit, devalued currency. Society's capacity to function as a whole is degraded, and its cohesion is eroded by the cultivation of an ideological climate, which opposes rigor and seeks ambiguity. The rules and criteria for rigorous thinking and clarity are lost. As clarity of speech fosters a climate of trust and honest trading, ambiguity opens the door to transgressions and lies.

For decades, there has been a trend of degrading traditional values and thus depreciating the words that confer meaning upon them.

In certain Western countries, talking about one's "nation" or "homeland" raises suspicions. The ideology that "national" and "social" are opposites tends to be embraced by public opinion.

Of course, one might well claim the concepts of "homeland" and "nation" have been exploited and used for self-serving ends by certain people. However, this is not a reason to consider these concepts dangerous or useless. In a healthy society, you do not simply throw away the concepts necessary to understand our historical development just because someone "polluted" them. You should simply make sure that you "clean them up," that you pass them on immaculate in the collective conscience. This is the modus operandi of societies that protect their legacies that do not allow them to go to the dogs.

In the twentieth century, we saw the depreciation of words such as "science," "intelligence," "politics," and "ideology." When I was a student in the 1960s, young people dreamed of becoming scientists because science was placed high on the scale of values. Science was salvation, it conveyed knowledge and all the innovations that changed people's lives. Young people also believed in the value of good; they dreamed of social justice and of a better future for our country.

This picture faltered and changed after the 1970s. The indiscriminate use of scientific results, pollution and environmental change, unemployment and poverty have a direct impact on the decline in the prestige of science.

The degradation of concepts is accompanied by the emergence of new ones. George Orwell wrote "if thought corrupts language, language can also corrupt thought." The syndrome of terminological design is becoming increasingly acute, where words with a "negative connotation" are replaced by others that are more neutral or politically correct. A blind person becomes, for the sake of political correction, "visually impaired" or "sight challenged."

The "wooden language" syndrome is well known, particularly in politics, which makes it difficult to find common ground to reach an understanding. This language is dangerous due to being misleading: saying one thing that might also mean something else, spreading hollow words, cultivating an ambiguity that is both complacent and appealing. Alan Greenspan, the former chair of the US Federal Reserve, was not known for being particularly clear in his speaking, and would ironically say, "I guess I should warn you, if I turn out to be particularly clear, you've probably misunderstood what I've said."

When a concept is lost or replaced by another, the very way in which we can conceive reality changes. Controlling and shaping language by creating new words or removing others ultimately makes it possible to control thought. Thus, a "defeat," a word with a troubling negative connotation, becomes a "tactical retreat," or a "mobile defense war." During a crisis, we enter a period of "negative growth," "economic rigor," or "economic austerity" leading to "rigor for modernization."

In recent years, we have seen an attempt at Orwellian redesign in the name of political correctness. We are told that in order to create a better world free of inequality and social injustice, we have to "clean up" our historical, scientific, and technical knowledge from references to individuals whose example and ideas can hurt and insult people who are discriminated against and live on the margins of society.

They thus suggest that any reference to giants of science such as Archimedes, Newton, Schrödinger, or Curie be banned when we teach their results. We are also told that curricula and the ways they are taught must change with a view toward more Diversity, Equity, and Inclusion—see [4] for an example.

The reasons for rejection vary either because it is not proper to indicate white superiority, or because people who lived in a particular place and time, for example, slave society, colonialist or fascist regimes, revolutions, and world wars, did not always make the "right choices."

Whatever the motivations of this movement, naive idealism or political self-interest, it is a dangerous trade in good sentiments. By purportedly seeking more justice, it aims to root out traditional values without actually helping to redress social injustice. It is downward leveling, a brainwash that wants to change the core values of our social system, such as individual values and excellence.

Only totalitarian regimes impose censorship and moralism in order to supposedly protect the common good. Language is man's most precious common tool for thinking and communicating.

The redesign or ostracism of concepts, self-censorship, and technology-assisted language control can imperceptibly change our "thinking software," the way system software is changed through over-the-air updates.

8.6.2 The Cycle of Lying

Lying marks the decline of societies and the moral impoverishment of individuals. I am, of course, talking about damaging lies, not white lies told to avoid offending or provoking someone.

The most common type of lie is the lie "in word," i.e., when you consciously say something different from what you know or believe to be true. You might know that something is "white" and say that it is "gray" or "black."

Another type of lie is the lie "in deed," i.e., when actions fall short of promises. You may have promised measures that will reduce prices and do exactly the opposite; you may hold a referendum supposedly to strengthen your country's negotiating power, while you are simply holding it to defuse a situation.

However, there is another type of lie that is far more dangerous and devious: the systemic lie. While the act of lying is conscious in the first two types, in the third type lying is incorporated, through automatic mechanisms, into the weak, faltering logic of the person telling the falsehood. This is what we call "fooling oneself." That is, we more or less automatically accept as true something which when you examine it soberly and rationally is completely wrong. It is a lie that charms us and tempts us toward an unrealistic, utopian impulse. We live through such experiences as children and adolescents. Considering our wishes to be reality, we create utopias, "myths" that are convenient to us.

What I want to discuss here is the establishment of this type of lie in our collective conscience and its disastrous consequences. In a healthy democracy, people telling falsehoods would be subject to massive and immediate rejection. Decline is inherent in a remarkable tolerance for lies. There are countless examples demonstrating that the triumph of the systemic falsehood makes it almost impossible to debate and criticize based on rational and ethical arguments.

I have explained that many mental functions are automatic. Mental rules and the ways we apply them are shaped through learning processes that function subconsciously. The way we use concepts determines their meaning. If in practice you court ambiguity and falsehoods, this becomes second nature. The meaning of the concepts slacks and the coordinates of thinking become blurred. As a result, the concepts of words slide away or expand as one sees fit.

Systemic lies lead directly to relativism, a creeping ideology that reigns supreme in declining societies. By mistreating and depreciating truth, systemic lies favor ambiguity, undermine relationships of trust and hinder any development of creative dialog in societies. I cannot fathom how ambiguity can be creative. It certainly helps "muddy the waters" and exact unconscionable contracts based on deliberately ambiguous wording of controversial points.

Some people are fond of the so-called "constructive ambiguity" as a means of reaching agreement between disagreeing parties by obfuscating thorny aspects of a debate. However, ambiguity should not be confused with abstraction as discussed in Sect. 4.1.2. Unlike ambiguity, abstraction is an absolutely creative act, a holistic way to break down complexity by using a filter. Confusing ambiguity and abstraction is like considering chicken scratches to be abstract painting.

When a person systemically lies, they consciously create myths and narratives that explain and justify their attitude. There are numerous examples of this syndrome of prevarication, which is intended to alleviate guilt and to drape self-interest with a mantle of superiority. Racism was based on theories of the superiority of certain races, while the barbarism of the Holy Inquisition invoked the justification that fire purifies the soul. You justify your intolerant attitude by espousing the theory that you are an ideologist and pure, and therefore your opponents are corrupt, traitors, collaborationists, etc.

Systemic lies distort discourse, cutting it off from reality, making it lose its creative power and role in shaping common visions. It cannot be embodied through purposeful action. It becomes empty rhetoric, pointless, boring, and even dangerous at times.

Systemic lies dull the moral sensitivity of people and blind them so that they cannot see reality soberly. People now resort to "myths" which are increasingly divorcing them from reality.

Lies become the "bad money that drives out good money," which ultimately dominates and excludes any realistic approach.

Let me conclude with a story where a grandchild regularly flatters his old grandfather, who is moved and gives him some pocket money. One day, the young boy overdid the praise, so much that his grandfather grumbled and, annoyed, told him: "I know you're lying, blast you, but what can I do, I like hearing the damn lies." We have learned to live with the lie and cannot do without it.

8.6.3 Self-Convenience and Lack of Self-Esteem

Have you wondered what your life would be like if you had a servant to satisfy your every whim and desire? If there were no obstacles, no difficulties, if you succeeded at everything effortlessly, if there was no stopping your appetites?

If humans had what satisfied their appetites automatically and effortlessly, they would be reduced to "lazy gluttons," merely a digestive tract.

If humans were born in a world without adversity, they would certainly be no more intelligent than plants. Moreover, their happiness would also be lesser to that of the plants because plants at least struggle to survive.

We live in a period of idolatry. We do not believe in our own forces; we consider any spiritual search to be unnecessary, we place our hopes for salvation solely in the material and ephemeral. We see the symptoms of the deterioration of ourselves, of our self-esteem. We seek outside support, as we do not have moral strength and fortitude.

The indigenous people once adored the conquistadors because they carried loud firearms that reminded them of lightning and thunder, which were the hallmarks of their god. At other times, societies were entranced by cars moving much faster than humans. Today, we have computers that solve problems more quickly than people. We are fascinated, without considering that cars, weapons, and computers would not exist without people, without human creativity.

After the 1970s, societies showed hypersensitivity to what can be considered a major effort to achieve difficult goals. Everyone talks about protecting everyone's rights, about "non-oppression," about avoiding stress. Few remember that there are also obligations and that as the saying goes, "no pain, no gain." As a result, we have seen a tendency in many European countries to ease educational programs, so that the "poor children" do not suffer, do not undergo traumatic experiences and, above all else, know and claim their rights. All this demagoguery is taken place with the assent of the state, the educational unions, and parents.

Making the bare minimum effort in every task has become the rule, sloppily patching everything up. We do not care about the future. The quality criteria that require rigor and precision in carrying out tasks are being slackened.

People only want to hear good news. They close their eyes to the important problems; they already face plenty, being overwhelmed by their daily complexities, and the demands of modern life. Let us be positive about everything, without complaining and caring about civic affairs, which will remain unresolved forever anyway.

In this downward spiral, only individuals and their wishes in the here and now count. People think nothing of future generations, they have no vision, they avoid difficulties and shut their eyes to the problems. This leads to societies that are tired and expect nothing from tomorrow. In contrast, dynamic societies establish visions and prepare the future, even if it means facing deprivation.

In the wake of the coronavirus crisis, I have seen how much people's egos have grown, especially among young people, who openly and shamelessly put their own entertainment before public health and the common good.

I have also been amazed by the demagoguery, the pathetic "understanding" shown by decision-makers and the media: "we understand, they are young people who want to have fun." As if it would be such unbearable torment to live through a summer with no rave parties and all manner of carousing—the way young people of my generation would spend the summer in the 1960s.

In conclusion, I would say that this leveling of the hierarchy of values is accompanied by a redefinition of concepts and the relationships between them.

Even hardened consciences need justification for non-compliance with the rules and their anti-social behavior. A very frequent attitude is the lack of a sense of responsibility and the excuse that "others are to blame" or it is because of general circumstances beyond our control.

This attitude is reinforced by the inadequate functioning of institutions, control, and evaluation mechanisms. It leads to a practical challenge to the "utility" of ethical principles and discourages efforts to achieve higher goals.

8.6.4 The Triumph of the Idiots

One aspect of social decline is the fact that mediocre personalities prevail in public life. The question is how social conditions influence and shape this situation.

It is obvious that the dearth of values and favoritism benefit the emergence of the idiots and discourage the integration of the capable. In a depraved system, the prevalence of idiots is a certainty. A capable-good individual will avoid such a system, seeking one that recognizes values, being confident in their abilities. On the other hand, an individual with reduced faculties of understanding and management will seek to integrate into a slackened system, which leaves room for irregularities and is not sufficiently vigilant.

What is an idiot? Someone who is unable to understand reality, to correctly interpret situations and to link them to actions.

It should be noted that this inability has nothing to do with one's knowledge and educational level, or even their IQ.

There are many types of idiots at every level of social organization across all communities. Useless idiots, the common numbskulls, account for the overwhelming majority and play a rather useful role in social organization. They serve as a type of "ballast" that causes friction in the system and makes it less sensitive to sudden changes.

Here I would like to talk about dangerous idiots who have managed to become part of a self-sustaining and relatively closed system of elites, who hold a country's political, military, and economic leadership positions.

One well-known story is that of Marie Antoinette who, when told that the French people were starving and had no bread to eat, responded "let them eat cake"!

I also recall a joke about an old, much-hated dictator who was on his deathbed. Crowds had gathered under his window waiting for his death to be announced in order to celebrate. Listening to the bustle of the crowd, the dictator was surprised and asked: "Why are all these people here?", and received the answer "To bid you farewell, my General." And he asked, "Well, where are they off to?".

Depending on the country and the era, the way in which elites arise varies. In the past, privilege was almost exclusively hereditary. In contemporary democracies, the elites are supplied by educational institutions that are open, as a priority, to young people of top classes—such as Harvard, Columbia, Yale, and Princeton in the USA.

Of all the types of idiots, the most harmful ones may be those who reached high places without even realizing what happened to them. When you obtain something painlessly, without fighting for it, then you are unworthy of respecting it and you often lose it quite quickly.

Throughout my life, I have had an innate interest in political affairs, particularly politics in France, where I have lived most of my life and in Greece, due to my origins. I am struck by the inadequacy of decision-makers at all levels—and particularly the fact that, despite their proven incompetence in managing public affairs, they are able not only to survive in public life but also to actually enjoy brilliant careers, with very few exceptions.

If you are a mismanaging grocer, your business will suffer losses and will soon drop off the face of the earth. If you are a mismanaging Minister, not only do you face no risk but, depending on the circumstances, you may actually be promoted. That is why political parties and governments contain an overwhelming number of idiots.

Of course, this phenomenon is universal and timeless. It is exacerbated by crises and occurs in every organization. Even in big technology companies, a good engineer will not be placed in a managerial position, while those who fail in general will be promoted; the adage "managers rise to the level of their incompetence" rings true.

Idiots are mistrustful by nature, especially of those who surpass them intellectually. When they do not understand something, they entrench themselves in predetermined views and do not even listen to their interlocutors. This has often happened to me when I have tried to explain issues that require a degree of concentration to ignorant managers.

A consequence of mistrust is instinctive cunning. Smart people are not cunning. They speak and act straight because they have confidence in their judgment and in their powers.

In order to maintain the acquis, dangerous idiots have certain defense mechanisms. First, they become systematically mistrustful, particularly of what exceeds their intellect, and especially of intelligent people, whom they avoid like the plague. An encounter with an intelligent person, however brief, can be very painful to them. One natural reaction is to reject them one way or another, to entrench themselves in derogatory or arrogant behavior.

One way in which idiots protect themselves is only enjoying the company of equivalent idiots, forming cliques of all kinds at work or in social life. A high-ranking idiot will never hire intelligent advisors. This results in a system that protects and promotes idiots, while rejecting any element that does not align itself with the system's logic.

A typical attitude of ignorant Ministers in the face of a difficult problem involving political cost is to set up a committee of experts to address the problem. As a rule, when the Minister's term of office ends, the committee has not even drawn any conclusions. There is a famous quote attributed to Talleyrand: "Ne rien faire mais bien le faire." It translates to "do nothing, but do it well."

Unfortunately, the prevalence of idiots is a symptom of the political, economic, and moral crisis that has plagued post-industrial societies. It undermines one of the foundations of democracy, which is meritocracy, a topic I will discuss in what follows.

8.7 About Democracy

The aim of a political system is to ensure economic prosperity and quality of life for its citizens. The demands for economic growth and social justice must not be made to compete with each other. Each should assist and complement the other. Social justice has an ethical, as well as a practical and technical dimension. The absence of formal and substantive justice is a long-term obstacle to the development of a healthy economy. When a country has a high rate of unemployment, when a segment of its population is barely surviving, then it is self-evident that the performance of the economy cannot be optimal. Democracy without economic progress is a utopia. Without financial resources, the popular verdict is but an empty shell, political will is but hot air.

It is common belief that democracy is the system that enables this dual purpose to be achieved in the best possible way: ensuring equality before the law and enabling every citizen to achieve self-actualization and, in particular, to create and develop their personality in conditions of safety and prosperity.

There is no perfect democracy. Despite the common formal characteristics, there are major differences in its essential modus operandi depending on each society and country. These differences stem from the degree to which citizens are committed to common values and, in general, from the effectiveness of institutions.

Democracy is not only formal rules—and this has been categorically proven by historical experience. It is well known that the popular verdict according to formally free, successive choices can lead a country to disastrous situations such as the emergence of Hitler's and Mussolini's dictatorships or the deep economic and social crises of the interwar period.

This system, which was originally conceived and tested in ancient Greece, can be considered to be optimal, but under certain conditions. These were the social conditions of the ancient city-states, which formed relatively small and well-structured groups. Citizens/individuals voluntarily surrendered their personal liberty, subjecting themselves to the idea of the common good in return for protection and solidarity. I will not further analyze the relations and links between individuals and society at large in ancient Greek cities.

Athenian society safeguarded its cohesiveness by directly or indirectly exercising strict control over citizens whenever they have broken away from certain standards. I am not referring to illegal acts, but simply behavior, which could, in a way, harm the cohesiveness of the whole and the principles of the political system—the institution of ostracism and the condemnation of Socrates are typical of this. It is also well known that the Athenian democracy did not forgive the mistakes of its leaders.

The ancient Greek model and the subsequent Roman *res publica* inspired modern democracies. Democracy is based on two principles.

One is equality, which takes three forms: *isopolity* (equal participation in joint decision-making), *isonomy* (equality before the law), and *isegory* (equality in expressing one's opinion).

The second equally important principle is meritocracy. That is, the best suited, the most worthy persons are those appointed to positions of responsibility and decision-making. Here I must clarify the difference between worthiness and excellence. Excellence is a distinction in a professional or scientific field. Worthy persons are those among the excellent who can provide important social or political work and have demonstrated practical virtue and skills. I believe the distinction is quite clear. It is a matter of competence, as well as of will to manage and contribute to public affairs. It is not achieved by obtaining diplomas and qualifications but through empirical testing. A distinguished scientist or entrepreneur may not be worthy of contributing to public affairs for reasons beyond their other capabilities, either because they prefer to be private citizens or because they are not so inclined.

Worthy people are elevated and acclaimed in practice. Of course, it stands to reason that the worthy would be chosen from among the excellent.

The first references to meritocracy can be found in Plato's *Republic*, where he describes an ideal hierarchical organization of society based on a classification of citizens. The subject is inexhaustible, and I am neither inclined nor best suited to make such analyses. I simply want to make it clear that equality alone is not enough and to stress how important the role of leaders and creators in such a system is.

I have already explained the role of institutions and values in modern societies. Democracy is the perfect form of government when every individual can responsibly and honestly manage their freedom—otherwise, it becomes a nightmare. In such cases, it should be replaced by a more centralized regime where at least decisions and actions are handled more effectively.

Laws often dictate what is forbidden and set the minimum obligations of individuals toward their fellow citizens and the state. However—and this cannot but be the case, unless you have a system where you "watch the watchmen"—it is impossible to control and impose a proper mode of operation when the sense of responsibility has become dull, when consciences have been corrupted.

The first and last bastion of democracy is the spirit of patriotism and the sense of responsibility among its citizens.

8.7.1 Meritocracy and the Role of Leaders

The application of principles of meritocracy requires the hierarchical structuring of power systems through institutions with appropriate planning and control mechanisms.

I would like to stress the importance of a hierarchical structure, which allows for centralized but efficient information management. For example, there are computer networks where, for technical reasons, all nodes are equivalent in decision-making; these are the so-called distributed systems, without a central computer to coordinate them. In such systems, the computational cost of coordination, for example, in order to reach consensus, is quite significant in comparison to centralized systems. For example, distributed systems are used by blockchain technology, which involves excessive computing and therefore energy costs to complete transactions.

The tumultuous history of revolutions and social change led to the establishment of commonly accepted democratic principles and structures. These are more or less well known and have been adopted by forms of government, which, at least in theory, adhere to the "democratic model." The problem is the proper application of these principles. In order to become a successful federation, it is not enough to copy the Swiss political system. The existence of proper structures is just a prerequisite. It is ultimately the behavior of citizens that determines the result.

Development and social well-being cannot be achieved without planning and organization that takes into account the current situation and aims at achieving long-term goals; in fact, more than goals, what is needed is a vision of what society wants to achieve and how to live in the distant future. For those who understand how organized forms of government work, such a vision cannot take shape through light "coffeehouse" debate.

I have lived in countries where there is a widespread perception that democracy is thereby guaranteed in abstract terms, through "broad participation," and through "collective processes." If it is true that the essence of democracy is broad consensus and agreement on a program of actions for the common good, it is not at all self-evident that this program must be drawn up through procedures involving the "base" (through general meetings and broad, leaderless working groups). I have seen such demagogic approaches meet with abject failure.

This idea can only lead to a hodgepodge, a ridiculous situation without logic and coherence. It often goes hand-in-hand with the idea that every member of the "base" holds a "piece of power," thanks to party, trade union, and other connections—and each person uses this piece of power to contribute to the shaping of choices, while, of course, defending their own well-intended interests. Thus, goals become an incongruous mishmash, the sum of the aspirations of the many, not a lofty yet realistic vision that the most worthy have drawn up in a rigorous manner.

In my life, I have had to participate in a large number of committees and councils, at every level, to formulate policies and evaluate structures. They have cost me endless hours of drudgery and tedious debates, with results ranging from negligible to completely counterproductive.

The sweeping concept that does not recognize the necessary role of leaders and assumes that everyone is equally worthy of governing, regardless of qualifications and competence, directly affects the very authority of democratic institutions. Unfortunately, governments are full of ignorant Ministers chosen according to party and other criteria, even though it is not difficult to find more suitable officials, even within the ruling party.

Such a concept is a violation of the principle that, in a democracy, all opinions should count equally when choosing between proposals. However, the process of formulating proposals must guarantee that they are technically consistent and contribute to the development of a realistic vision for the common good. Visions cannot spring forth from endless discussions and upheavals. They are the creations of inspired minds who take the lead, make a proposal and point the way forward. They are thought up by people with recognized abilities to create and guide.

I would also like to emphasize that in Greek the word "*demiourgos*," which means "creator," or "maker" and is the root of the English word demiurge, is a compound of "*δῆμος*," demos (populace of a democracy as a political unit) and "*ἔργον*," érgon (work), and initially meant a man who practices a profession. Later, the word came to mean someone who carries out a "public-interest" work. It was a title conferred upon officials in many ancient city-states who were responsible for organizing public affairs. Finally, in Platonic philosophy, the word took on the meaning of the Maker or Creator of the world.

Of course, this gives rise to the following question: how do leaders emerge? How do we put the right people in the right position? I would answer that the problem is not how to find them, but how to convince them to engage with public affairs. There are worthy people who are willing and able to contribute to the common good. However, they must be provided with guarantees that their opinion will be respected and that their name will not be used for self-serving purposes.

It is commonly believed that the wider the process of selecting decision-makers for a position, the better the result will be. This is another moronic criterion that ignores the substance, namely the clear definition of the qualifications required and the fair evaluation of the candidates.

At times, I have conducted rough opinion polls of who are best placed to govern in Greece and France, two countries I know quite well. Asking people with common sense from various walks of life who the most capable in a government are, you see that there is an astounding consensus regarding the persons named. It does not take a genius to understand that so-and-so is not quick on the uptake, that another person is susceptible to suggestion, and that yet another has no desire to take risks in order to do something meaningful. In other words, you need not be clairvoyant in order to know. One might well ask if it is so simple, why capable and honorable people are rare in governments?

The answer to this question is also simple. First, no single party consists of enough people of demonstrable worth to form a good government. Second, by their very nature, political parties must serve all trends, inevitably including the ignorant, the corrupt, etc. in their ranks. The leader of the ruling majority often knows that they do not have the best possible partners, but they do not have a choice.

One way of finding the worthy is by not choosing the most demonstrably unworthy in the first place, as they do not meet at least one of the two requirements: integrity and recognized competence. Such a purge makes the game of choice much simpler. Fortunately, worthy people still exist—but unfortunately, they stay far away from public affairs.

When I was young and radicalized in France in the early 1970s, I had been impressed by Charles de Gaulle, despite myopically rejecting him as being "conservative." The General was famous because he did not pull any punches with the

French. When he addressed them on television, he would call a spade a spade and ignore the political cost. This man, who led the resistance against the Germans during World War II, laid the foundations for an independent France by creating a vision for its decolonization and its development as an industrial and military force.

As a mature researcher, I had the opportunity to learn first-hand, through conversations with famous French physicists, how de Gaulle made France a nuclear power. He secretly had the first atomic bomb built by ignoring not only the reactions of France's allies but also the opposition from the majority of the country's political establishment. What I mean to say is that the action of a handful of enlightened, worthy people can change the course of a country's history.

I want to stress the role of the human factor in creating, envisioning, and planning the future.

In Western democracies, visions are drawn up by organized groups tasked with an institutional role (lobbies, think tanks, research institutions, and academies). There are constants in the policies of major states, regardless of who is in power at any given time. This is the case with the USA, as well as China, Germany, and many other major players on the international political scene, such as Israel, India, Japan, and Korea. We can judge and criticize the way decisions are made, who they benefit, etc. However, foresight and planning are better than abandonment and inertia.

In a country such as Germany, conflicts die down when the future of the country is at stake. Let us recall that Schröder's Social Democratic government took strict austerity measures in order to address the impending crisis, with no regard for the political cost.

In Germany, industrial strength is based on sectors such as the automotive industry, the chemical industry, and production automation. Industry representatives and research institutions play a key role in identifying priorities per sector. They work with every German government to develop innovative programs, for example, for electric cars or new forms of energy. At the same time, there are institutions such as non-profit organizations, think tanks and experts, as well as press and media organizations whose words carry weight.

I remember in the past, in France, what effect an article in the newspaper *Le Monde* would have when it criticized the government or when it provided guidelines for the country's foreign policy.

How do the strategies of the USA for each sector take shape? The system has organized lobbies that make proposals and at times take initiatives independently of the federal government. The technological supremacy of the USA and the strategies to support it are shaped by institutions through group discussions/conflicts where the military, business, academics, bankers, etc. are represented and are relatively independent of the political context at any given time.

I could, of course, also talk about other countries, such as China, which has a centralized planning model with well-defined priorities that mobilizes all the resources needed to realize them, aiming at national independence and self-sustaining growth above all else.

I will conclude by saying that there is a distorted view of what a democracy is; it emphasizes the role of the popular verdict but ignores how essential the role of leaders and creators is in order for the state to function properly.

To paraphrase a well-known aphorism of Albert Einstein, I would say that "democracy without meritocracy is blind—democracy without equality is lame."

8.7.2 Corruption and Bureaucracy

The scourge of modern democracies is bureaucracy, a system where the most serious political and administrative decisions are made by appointed state officials. This system of administration, which has its roots in ancient Egypt and ancient China, aims at the self-serving control of administrative structures. It imposes increased complexity on procedures, which leads to inefficiencies in the management of the state. Today, it is considered a symptom of ignorance and corruption on the part of those in power.

Depending on the circumstances, bureaucracy may interfere with government functions for self-serving purposes or even become dangerous for individual liberties. It can thus cover its impotence to handle public affairs and can organize networks of corruption for its own benefit.

The biggest problem of an incompetent administrator is how to remain in office without stirring the pot, making decisions on thorny issues. An administration that does not want to decide on a problem will look for pretexts by creating structures, committees that will write long reports. This will prolong the procedures and create relationships of dependence for those seeking a solution.

In corrupt regimes, a large and poorly organized administration hinders seamless operation by requiring appropriate tolls at every level of access to services. It is paradoxical that bureaucracy is often set up on the pretext of ensuring better and more efficient handling of affairs. However, if we were to believe this argument, we must realize that even the best decision is useless when it is made too late or when the cost of the decision-making process is excessive.

In many cases, the setting up of committees—particularly populous and, as they refer to themselves, "representative" ones, with trade unions, organizations, etc.—is seen as a guarantee of seriousness and quality. There is a tragic misunderstanding here about how creative ideas and solutions are born.

Large committees are definitely a problem that negatively affects creativity and efficiency. While it is right that a proposal, once formulated, be handed over to the many for criticism, it is not the many who will make a proper proposal. Creation is a solitary action. Creators need self-concentration to mobilize their ingenuity and knowledge in order to analyze a problem and come up with solutions.

Bureaucracy usually issues orders, both to experts and non-experts, to come up with long analyses and reports, purportedly to help it solve a problem. I have seen such voluminous works that either often draw obvious conclusions or are so poorly written that no clear conclusions can be drawn at all.

In my career as a researcher, having occasionally taken up administrative positions to manage research and innovation, I have had the opportunity to study relevant analyses carried out by well-paid consultancy firms. I generally found them to be superficial, limiting themselves to describing situations and reaching painless and obvious conclusions. Analysts will never risk losing a paying customer by voicing the big truths that could offend their commissioner.

I have experienced many cases of bureaucracy.

One illustrious example of bureaucratic organization is the European Commission, where the complexity of the structures competes only with the complexity of its regulations. The perennial aim is, supposedly, to respect the rules and to manage European funds properly. What everyone disregards, though, is that control mechanisms only make sense if the cost of the audit process is commensurable to the resources being managed.

I have seen endless procedures to audit pennies. There is a different way to achieve proper management with lower control costs: when the penalties are severe in any case of violation. This shockingly simple observation eludes the Eurocrats. What is needed is fewer auditors and tighter control according to simple rules for meeting substantive obligations and commitments.

We have not yet fully understood the informational nature of human interactions. I remember in primary school the teacher who taught us the "rule of three" using the following example for inversely proportional amounts. "It takes 10 people 10 days to dig a field. How many days does it take 20 people?". Proportionality implies that 20 people can work in parallel (independently), without each person hindering the other. It applies when there is perfect allocation of tasks, with minimal interactions.

The same question, when asked with regard to collaborative work such as a software program to be written by a team of engineers, would make no sense. More engineers, poorly coordinated, will not shorten but actually lengthen the time needed for completion [5].

Cooperation involves a cost, and we need to prove that the gains in productivity and quality are greater than the cost of coordination. Pyramid structures are certainly more efficient than completely flat ones with no central coordinating body.

If I increase the number of computers I use to solve a problem, the gain in speed of resolution does not necessarily increase depending on the number of machines. In fact, beyond a certain threshold, I may not have any increase at all. This is what it means when we say that we do not have scalability.

What happens with computers also happens with people to a great extent—and the problem is compounded by the fact that people can demonstrate selfish behaviors that can make the game of cooperation even more complex.

Those in power have a natural tendency to use even a small portion of that power to maintain their position within the system. The fight against bureaucracy must be a constant concern for every democracy.

References

1. https://en.wikipedia.org/wiki/Precautionary_principle
2. https://en.wikipedia.org/wiki/Boeing_737_MAX_certification
3. Philip Mirowski, *Privatizing American Science*, 2011, Harvard University Press, ISBN 9780674046467
4. https://poorvucenter.yale.edu/Antiracist-Pedagogy (accessed on 08/01/22)
5. https://en.wikipedia.org/wiki/The_Mythical_Man-Month

Chapter 9
Epilogue

The book is the crystallization of thoughts I have had over many years, spurred on by an innate curiosity regarding the big questions for which I did not find satisfactory answers. In this quest, I have been helped by my foundations in applied logic and my knowledge as a computer scientist and engineer. I have faithfully adhered to one rule: to distinguish what I can speak about with confidence and to define the boundaries of knowledge as clearly as possible.

That is why my first concern was to determine which of the big questions can be studied in a rigorous manner, and which cannot. I soon discovered that there is indeed a very clear distinction, which arises from the logical nature of the problems, and which everyone should understand. This would greatly limit vapid discussions, with the pretext of "scientificity," concerning ontological or teleological issues.

I therefore aimed at studying gnoseological problems relating to the development of knowledge—particularly scientific knowledge—and the application of knowledge—particularly technical knowledge. Characterizing knowledge as information directly connects it to informatics and computers.

I have explained that information is an entity separate from the matter/energy of the physical world. The proposed framework relies on a dualist vision of the world: on the one hand, the physical world, and on the other hand, information and knowledge as a creation of the human mind. This approach transcends the impasses of an approach that considers matter/energy as the fundamental entity and information as an emergent property of matter. How can we comprehend the properties of our thinking by studying physical phenomena alone?

This outlook helps us understand the close interaction in the development of scientific and technical knowledge. Unfortunately, we overemphasize the importance of science while sidelining technical knowledge. It is a shared view that curricula focus on accumulating knowledge through rote learning, and do not give due consideration to applications and creativity.

I rank informatics alongside the key fields of knowledge, together with physics, biology, and mathematics, those fields that provide basic cognitive tools to all other fields.

J. Sifakis, *Understanding and Changing the World*,
https://doi.org/10.1007/978-981-19-1932-9_9

Languages, both natural and non-natural, provide the "toolbox" to describe what we observe and to understand the world. I have explained that the methods we use to break down complexity have limits—and that ignorance of those limits leads to over-optimism and over-expectations. I have analyzed complexity factors for the predictability of phenomena and the buildability of artifacts. Thanks to the use of computers, we can overcome certain obstacles and broaden the horizons of knowledge.

The relationship between natural and artificial intelligence is a very topical issue, particularly with the discovery and widespread use of learning techniques to develop smart systems. What will be the equilibrium point between autonomous systems and human control over their functions in the future? This will depend on technological progress, as well as the political choices that societies will make in order to avoid a kind of alienation from decision-making that would result from unrestricted automation.

Can we fully trust the autonomous management of critical resources and services? One thing is for certain: after *Homo sapiens*, humans who make tools that multiply their muscular power and create artifacts, we will see a species of humans who, thanks to the use of computers, will multiply the possibilities to control and change the world. However, the question is whether this "boom" in intelligence will be accompanied by a commensurate development of collective consciousness regarding its proper management.

Informatics provides a basis for comparing natural and artificial intelligence, as well as possible synergies between them. I have proposed a "mechanistic" view of consciousness and of the management of freedom, which highlights the complexity of the problem of understanding mental functions and their simulation by machines. It leads to a study of human social behavior, as defined not just by material but also—and mainly—by information and communication factors.

I have tried to demonstrate the informational character of social organizations by analyzing the mechanisms that shape values, the processes for creating and reallocating resources, the management and development of knowledge. The decline of societies can be understood as an enervation of structures and institutions that allow for a stable, commonly accepted value system and ensure the effective and secure circulation of information and knowledge.

We can gain a deeper understanding of organizational principles in light of the theory of computation. A pyramidal organization based on meritocracy is the only form of organization that allows for effective planning that enables the most worthy, the creators, to develop visions and proposals for the future and submit them to the judgment of the public.

Equality takes on its full significance when all citizens have equal opportunities to make informed choices and equal opportunities to climb up the pyramid. Meritocracy does not cancel the idea of equality but, on the contrary, it makes it more meaningful.

While writing this book, I adhered to a technical approach and, at the same time, tried to explain its limitations. I have no doubt that new knowledge will allow people to make progress in this game of understanding and changing the world.

However, caution is necessary here. Just as an airplane is not a bird, a computer is not a human mind.

I find it hard to imagine how creations can exceed the intelligence of their creators. However, it is possible that the creators end up dominated by their creations, either due to being unable to manage their complexity or due to intellectual laziness and a propensity for freedom from the burden of choosing responsibly—and that is a terrifying scenario that I would prefer not to imagine at all.

Index

J. Sifakis, *Understanding and Changing the World*,
https://doi.org/10.1007/978-981-19-1932-9

Printed in the United States
by Baker & Taylor Publisher Services